Springer Series on

Atoms+Plasmas 23

Editor: I. I. Sobel'man

Springer
Berlin
Heidelberg
New York
Barcelona
Budapest
Hong Kong
London
Milan
Paris
Singapore
Tokyo

Springer Series on

Atoms+Plasmas

Editors: G. Ecker P. Lambropoulos I. I. Sobel'man H. Walther

Managing Editor: H. K.V. Lotsch

1 **Polarized Electrons** 2nd Edition
By J. Kessler

2 **Multiphoton Processes**
Editors: P. Lambropoulos and S. J. Smith

3 **Atomic Many-Body Theory**
2nd Edition
By I. Lindgren and J. Morrison

4 **Elementary Processes in Hydrogen-Helium Plasmas**
Cross Sections and Reaction Rate Coefficients
By R. K. Janev, W. D. Langer, K. Evans Jr., and D. E. Post Jr.

5 **Pulsed Electrical Discharge in Vacuum**
By G. A. Mesyats and D. I. Proskurovsky

6 **Atomic and Molecular Spectroscopy**
2nd Edition
Basic Aspects and Practical Applications
By S. Svanberg

7 **Interference of Atomic States**
By E. B. Alexandrov, M. P. Chaika and G. I. Khvostenko

8 **Plasma Physics** 2nd Edition
Basic Theory with Fusion Applications
By K. Nishikawa and M. Wakatani

9 **Plasma Spectroscopy**
The Influence of Microwave and Laser Fields
By E. Oks

10 **Film Deposition by Plasma Techniques**
By M. Konuma

11 **Resonance Phenomena in Electron-Atom Collisions**
By V. I. Lengyel, V. T. Navrotsky and E. P. Sabad

12 **Atomic Spectra and Radiative Transitions** 2nd Edition
By I. I. Sobel'man

13 **Multiphoton Processes in Atoms**
By N. B. Delone and V. P. Krainov

14 **Atoms in Plasmas**
By V. S. Lisitsa

15 **Excitation of Atoms and Broadening of Spectral Lines**
2nd Edition
By I. I. Sobel'man, L. Vainshtein, and E. Yukov

16 **Reference Data on Multicharged Ions**
By V. G. Pal'chikov and V. P. Shevelko

17 **Lectures on Non-linear Plasm Kinetics**
By V. N. Tsytovich

18 **Atoms and Their Spectroscopic Properties**
By V. P. Shevelko

19 **X-Ray Radiation of Highly Charged Ions**
By H. F. Beyer, H.-J. Kluge, and V. P. Shevelko

20 **Electron Emission in Heavy Ion–Atom Collision**
By N. Stolterfoht, R. D. DuBois, and R. D. Rivarola

21 **Molecules and Their Spectroscopic Properties**
By S. V. Khristenko, A. I. Maslov, and V. P. Shevelko

22 **Physics of Highly Excited Atoms and Ions**
By V. S. Lebedev and I. L. Beigman

23 **Atomic Multielectron Processes**
By V. P. Shevelko and H. Tawara

24 **Guided Wave Produced Plasmas**
By H. Schlüter, Y. M. Aliev, and A. Shivarova

25 **Quantum Statistics of Strongly Coupled Plasmas**
By D. Kremp, W. Kraeft, and M. Schlanges

Viatcheslav Shevelko
Hiro Tawara

Atomic Multielectron Processes

With 110 Figures and 17 Tables

Springer

Dr. Viatcheslav Shevelko
P.N. Lebedev Physical Institute
Leninskii pr. 53, 117924 Moscow
Russia

Prof. Hiro Tawara
National Institute for Fusion Science
Toki 509-52
Japan

Library of Congress Cataloging-in-Publication Data

Shevelko, V. P. (Viacheslav Petrovich) Atomic multielectron processes / Viatcheslav Shevelko, Hiro Tawara. p. cm. – (Springer series on atoms + plasmas; 23) Includes bibliographical references. ISBN 3-540-64235-8 (Berlin: alk. paper) 1. Collisions (Nuclear physics) 2. Electronic excitation. 3. Atomic transition probabilities. I. Tawara, H. (Hiroyuki), 1936– . II. Title. III. Series. QC794.6.C6S49 1998 539.7'57–dc21 98-30693 CIP

ISSN 0177-6495
ISBN 3-540-64235-8 Springer-Verlag Berlin Heidelberg New York

Cover concept: eStudio Calamar Steinen
Cover production: *design & production* GmbH; Heidelberg
Typesetting: Steingraeber, Heidelberg
SPIN 10673279 55/3142 - 5 4 3 2 1 0 - Printed on acid-free paper

Preface

Multielectron processes occurring in collisions of atoms, ions and molecules with photons and charged particles belong to one of the most interesting and challenging areas of the modern physics of atomic and electronic collisions. While single-electron processes associated with transitions of one electron in the target or projectile are well understood conceptually and mathematically, a basic understanding and study of many-electron processes require the use of highly advanced experimental techniques and sophisticated theoretical methods which should come out of the framework of the traditional one-electron approximations.

In the present monograph, we will consider such processes as multielectron excitation, multiple ionization, multiple-electron transfer, excitation with simultaneous ionization, multielectron photoionization arising in a single collision of an atomic particle with a photon, electron or another charged particle. These collisions take place via various multielectron channels such as multiple excitation or ionization of the outermost electrons or a single inner-shell excitation or ionization accompanied by relaxation of many-electron Auger or radiative processes.

Multielectron processes are very interesting in understanding the nature of the many-electron and many-particle transitions. On the other hand, under certain conditions, the effective cross sections of these processes can be quite large and reach up to 30–50% of the total cross sections. They play a quite significant role in different physical applications: modeling of plasma processes, charge-state evolution of atoms exposed to an electron beam, projectile charge-changing reactions responsible for the beam lifetimes of stored ions in acceleration devices such as a storage ring or a synchrotron.

Multiple-electron transitions are basically many-body phenomena depending significantly on the correlation effects between electrons involved. This fact is in a sharp contrast to the single-electron transitions where electron-electron correlations play, as a rule, a minor role. The term 'correlation' usually refers to the correlation effects between electrons caused by the Coulomb interaction, however, other kinds of correlation, e.g., time or spin correlations, are also possible.

Effective cross sections (or probabilities) of the multielectron atomic processes may be very large. Here are a few examples. The experimental double- and single-ionization cross sections of Mg atoms by electrons with the energies $E \approx 1\,\mathrm{keV}$ are of comparable size: $\sigma_2(E) \approx \sigma_1(E) \approx 10^{-17}\,\mathrm{cm}^2$. This can be understood due to a significant contribution of the inner-shell electron ionization followed by autoionization processes. Experimental cross section for an ionization of $m = 30$ electrons from a Xe atom in collisions with a 15.5 MeV/u U^{75+} ion is about $\sigma_{30} \approx 10^{-18}\,\mathrm{cm}^2$, compared with the single-electron ionization cross section of $10^{-14}\,\mathrm{cm}^2$. Here, the involvement of the inner-shell ionization processes followed by a series of cascades is evident. The maximum four-electron photoionization cross section of Xe at the photon energy $\hbar\omega \approx 160\,\mathrm{eV}$ is $\sigma_4(\omega) \approx 3.2 \times 10^{-19}\,\mathrm{cm}^2$. Therefore, multielectron processes play an important role in elementary atomic reactions and should be properly accounted for in analysis of the collision processes, e.g., in calculations of the beam attenuation kinetics.

At present, multielectron processes are more or less understood at incident ion energies higher than those corresponding to the electron-orbital velocities. They can be reasonably well described in terms of the independent one-electron processes (so-called Independent Particle Model, IPM). The role of multielectron processes significantly increases with the projectile and target charges and the relative velocity as well. More precisely, the ratio of the projectile charge to the relative velocity is known to be the important parameter. For example, the cross sections of multiple-electron capture at low-collision energies have been found to be large. In 4 keV/u Xe^{27+}+Xe collisions, the process $\mathrm{Xe}^{27+} + \mathrm{Xe} \rightarrow (\mathrm{Xe}^{11+})^* + \mathrm{Xe}^{16+} \rightarrow \mathrm{Xe}^{22+} + \mathrm{Xe}^{16+} + 11e^-$ has been observed, indicating that 16 electrons are captured into the projectile which, in turn, is stabilized through 11 cascades of the autoionization. Such multiple-electron processes are presently out of accurate theoretical treatments.

The present book is aiming at the first comprehensive collection of the cross sections for many-electron processes together with the physical background including dependences on the relative velocity, the projectile and target charges, transition energies, the number of the ejected electrons and other atomic parameters. The book comprises a series of tables, figures and empirical formulas which can be used conveniently for students and specialists working in the fields of atomic and plasma physics, chemical and laser physics and in a number of the applied sciences.

We would like to express our sincere gratitude to all colleagues who helped us in preparing this manuscript, especially, C.L. Cocke, B. DePaola, A. Müller, N.V. Novikov, L.P. Presnyakov, P. Richard, R.D. Rivarola, E. Salzborn, V.A. Sidorovich, N. Stolterfoht, Th. Stöhlker, J. Ullrich, D.B. Uskov and T.J.M. Zouros.

Moscow-Toki, Summer 1998 *H. Tawara, V. Shevelko*

Glossary of Terms

Fundamental Atomic Constants

The system of the atomic units (a.u.) is used: $e^2 = m_e = \hbar = 1$.

$a_0 = \hbar^2/m_e e^2$ $= 0.529\ 177\ 249(24) \times 10^{-8}$ cm	Bohr radius
$E_0 = e^2/a_0 = 27.211\ 3961(81)$ eV $= 2\ Ry$	Energy
$1\ Ry = m_e e^4/2\hbar^2$ $= 13.605\ 6981(40)$ eV	Rydberg energy
$\tau_0 = \hbar^3/m_e e^4 = a_0/v_0$ $= 2.418\ 884\ 33(11) \times 10^{-17}$ s	Time
$v_0 = e^2/\hbar = 2.187\ 691\ 417(98) \times 10^8$ cm/s	Velocity
$\pi a_0^2 = 0.879\ 735\ 6696(80) \times 10^{-16}$ cm^2	Cross section
$\alpha = e^2/\hbar c$ $= 1/137.035\ 9895(61)$	Fine-structure constant
$c = 1/\alpha = 2.997\ 924\ 58 \times 10^{10}$ cm/s	Velocity of light in vacuum

The values given above are extracted from the set of constants recommended for international use by the Committee on Data for Science and Technology (CODATA), [E.R. Cohen and B.N. Taylor: *The Fundamental Physical Constants*, Phys. Today **48**, Part 2 (August 1995)].

List of Symbols

A^a	Autoionization transition probability
A^r	Radiative transition probability
B	Branching-ratio coefficient
E	Projectile energy
E_{cm}	Center-of-mass energy
f	Distribution function
I	Binding energy; ionization potential
I_m	Ionization energy to remove m electrons
K	Momentum transfer
ℓ	Orbital quantum number
m	Number of ejected electrons
n	Principal quantum number
N	Total number of the target electrons
P_s	Probability for single-electron process
R_m	Ratio of $m-$fold to single-ionization cross sections
TI	Transfer ionization
TS	Two-step process
v	Relative velocity
v_p	Projectile velocity
X^{q+}	Ion with charge q
Z	Nuclear charge
Z_p	Projectile charge
Z_T	Target nuclear charge
β	Electric polarizability
λ	Wavelength
ρ	Impact parameter
σ_m	m-fold ionization cross section
σ_+	Net ionization cross section
θ	Scattering angle
ω	Photon frequency

Table of Contents

1. Basic Experimental Techniques

This chapter starts with a description of the various important and requisite experimental techniques to study the multielectron processes including the electron capture, excitation and ionization in ion–atom, electron–ion and ion–ion collisions.

1.1 Targets

Most experiments have been performed with the ground-state projectile particles and the ground-state target species of both atoms and molecules [1.1, 2]. Very few investigations have been reported for the state-selective $(n\ell)$-specified projectile ion collisions where n and ℓ are the principal and orbital quantum numbers, respectively. The limited information of the cross sections has indicated that even a very small fraction (a few percent) of the excited species in the projectile beams or targets can change the observed cross sections significantly. However, the quantitative investigations involving the excited target species are relatively new and still limited [1.3].

1.1.1 Ground-State Species

Thermal energy targets. The ground-state targets can be prepared simply in the form of thermal gases or by evaporating proper liquid or solid materials. For details, the readers are referred to [1.1, 2]. Sometimes, the neutralized beam targets are used when they can not be prepared in a usual way. In such cases, first, the ions are produced in the ion source accelerated to a proper energy, and after mass-charge selection, they capture an electron while passing through gases or some other media including solids and surfaces, and are neutralized thus forming the neutral atoms. It should be noticed that such neutralized atoms contain a significant fraction of excited or metastable state atoms since the electron is often captured into the excited state of the ion. It is, in principle, possible to minimize their fraction by choosing the neutralizing target atoms or molecules in such a way that the charge-transfer processes are resonant in forming the ground-state species.

It is expected that this can be avoided when the negative ions are used to form the neutral atoms through electron *stripping (detachment)* where the most loosely bound electron is likely to be removed from the negative ions, instead of the electron capture into the positive ions [1.4]. Indeed, by now it has been confirmed that most of the elements in the periodic table, except for the rare-gas atoms, can form negative ions with sufficient intensities. The most important difference is the density in two types of targets which are different by many orders of magnitude. The gas pressure of atom or molecule target can be controlled over a wide range from 10^{-5} up to 10 Torr meanwhile the equivalent pressure of the neutralized beam is usually only of the order of 10^{-10} Torr or less. Another difficulty arises in formation of atomic targets from some molecules. For example, atomic targets are generated through dissociation of molecules at high-temperature ovens (made from tungsten for H_2 and rhenium for O_2) or electrical discharge processes where the accurate information of the fraction of the dissociation rates is required.

Ultracold gas targets. Such thermal energy targets are known to limit the observed energy resolutions. Systematic investigations of a number of collision reactions, such as the multiple-ionization processes, need a high-energy resolution spectrometer in determining the momentum of both slow recoil (secondary) target ions and secondary electrons (Sect. 1.5.3). For this purpose, the thermal energy of the target atoms becomes a dominant obstacle in order to get a high resolution.

Recently, a new technique has been developed [1.5] (Fig. 1.1). There, the target-gas atoms are cooled down to 14 K using the cold head of an ordinary cryo-pump and, after effusing, form a localized supersonic target with small transversal velocities which is located inside the collision region. Presently, this cold-target system is widely used in high-resolution measurements involving helium targets which provide a lot of new information on the collisions both at high and at low collision energies.

1.1.2 State-Selected Excited Target Species

Alkalis. Alkali and alkaline-earth metal atoms in the excited states (not necessarily in the metastable states) can be generated through the laser excitation [1.6–8]. For example, the ground-state $3s$ electron in Na can be easily excited to the $^3P_{1/2,3/2}$ state by the laser beam. Similarly, Rb atoms can be formed in their state-specified excited states as high as $n = 10$–20 by lasers (Fig. 1.2). It is also important to remember that the linearly or circularly polarized atoms can be obtained through the polarized laser beam as well [1.9, 10]. Except for these elements, other atoms can not be excited easily by lasers as the modern laser energy is still not sufficiently high. In

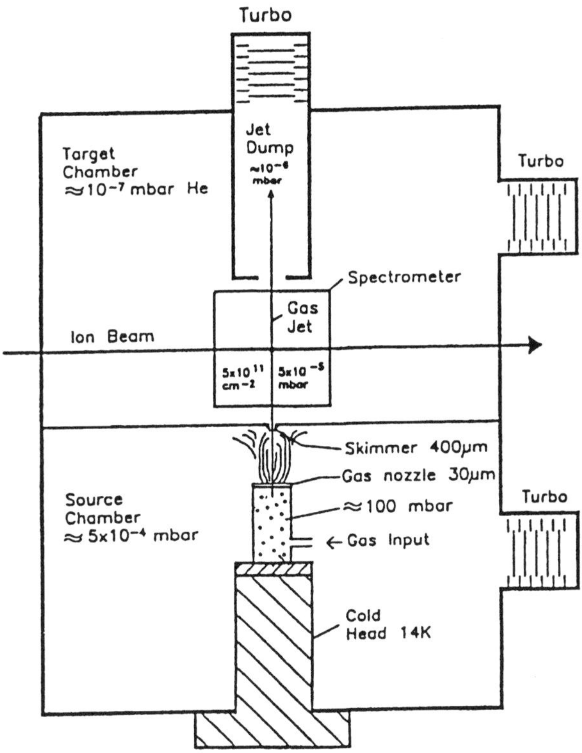

Fig. 1.1. A scheme of a cold-gas target where the gas atoms are cooled down while going through a cold head (at the bottom of the figure) at 14 K. The effective He densities at the collision region is about 5×10^{-5} Torr equivalent to the target thickness of 10^{11} atoms/cm^2. The projectile ions cross the cold target at the center from left to right. From [1.5]

principle, multiphoton excitation processes of some elements using ultrahigh power lasers are possible.

Other elements. Some target atoms can be formed through the electron capture by ions as mentioned above. For example, it is known that roughly 30% of He^0 $(1s2s)^3S$ metastable-state beam can be obtained through such electron-capture processes by $\mathrm{He}^+(1s)$ ions passing under proper target gases though they are not in a pure form but in a mixed form containing the ground-state beam as well. Accurate knowledge of the fractions of the excited species is required. Similar techniques can be applied in order to get the charged species.

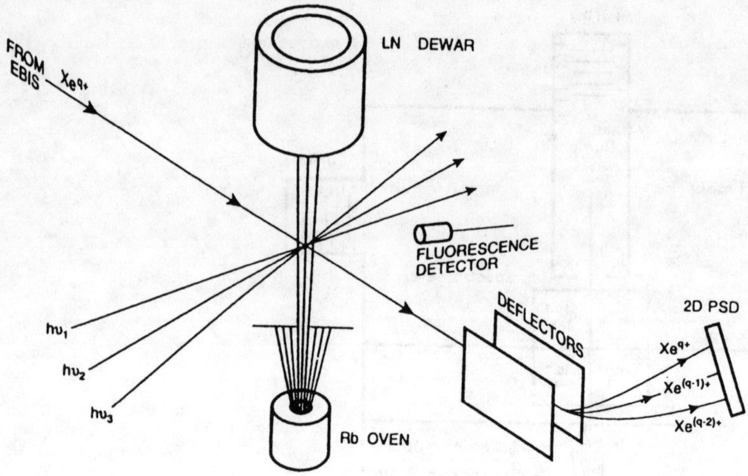

Fig. 1.2. A scheme of laser-excited Rb($n\ell = 10f$) atom targets. The ground-state Rb atoms are provided through a heated oven and excited by three lasers. The three-step excitation via lasers is used. The first laser excites Rb via $5s \to 5p$, the second $5p \to 4d$ and finally the third $4d \to 10f$ transitions. Xe^{q+} ions from EBIS (Sect. 1.2.2) cross the Rb target atoms which are trapped on the dewar surface cooled down at liquid-nitrogen temperatures (LN Dewar). After collision, the projectile Xe ions are charge-analyzed with a pair of the electrostatic deflectors and finally detected with a two-dimensional position-sensitive detector (2D PSD). The intensities of the lasers are monitored with a fluorescence detector. From [1.7]

1.2 Electron and Projectile Ion Beams

The formation techniques of the controlled electron beams are well known. Still a simple hair-pin type electron gun can provide the electron beams with resonable intensities. Also a powerful electron beam with intensity of 500 mA has been developed for ion collision experiments.

With the recent advance of ion source technologies, ions with different charge states up to U^{92+} ions can be produced directly in some ion source. The accelerators of various types combined can produce ions over a wide range of collision energy up to TeV.

The negative ions of most the elements in the periodic table, except for rare gas and some other atoms, have been known to be produced under proper surface conditions.

1.2.1 Electron Beam Sources

In a number of collision experiments with neutral-gas targets, relatively simple electron guns can provide sufficiently high-density electron beams.

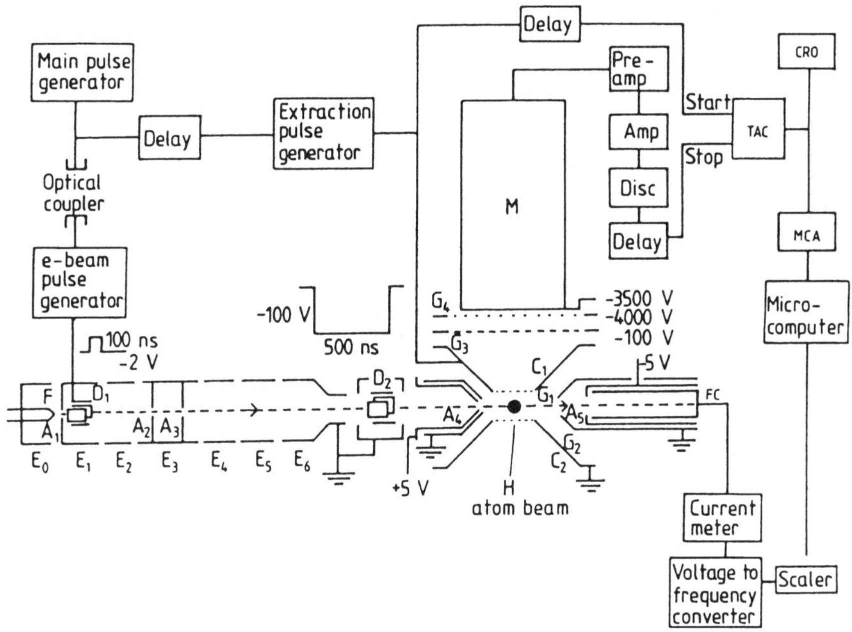

Fig. 1.3. A pulsed electron-neutral gas target collision system. The electron beam is chopped after extraction through an anode, then is sent into a collision region and finally arrives at a Faraday cup. Immediately after the electron beam has passed through the collision region, the pulsed extraction field between two electrodes C_1 and C_2 is applied and the extracted ions are accelerated into a detector. From [1.11]

Most of recent electron-impact multiple-ionization experiments involving relative intensity measurements of the multiply charged ions produced from the thermal neutral gases are based upon the pulsed electron beam, instead of the steady electron beam (Fig. 1.3), in combination with the jet gas targets [1.11]. A pulsed electron beam is sent to cross the ground-state atomic beam thermal (jet) target. Immediately after these electrons pass the interaction region, a pulsed electro-static field is applied to extract the secondary ions produced in collisions with the incident electrons which arrive at a particle detector such as an electron multiplier. Then, the time-of-flight analysis gives the information of the charge state and mass of the secondary ions. Generally, this technique can provide reasonably accurate relative cross sections but some normalization to the known cross sections is necessary to get the absolute partial cross sections. In these experiments, a relatively weak ($< 100\,\mathrm{nA}$) beam is sufficient to perform some counting experiments.

But, as will be mentioned later, the crossed electron–ion beams experiments need much more intense electron beams. For example, very powerful

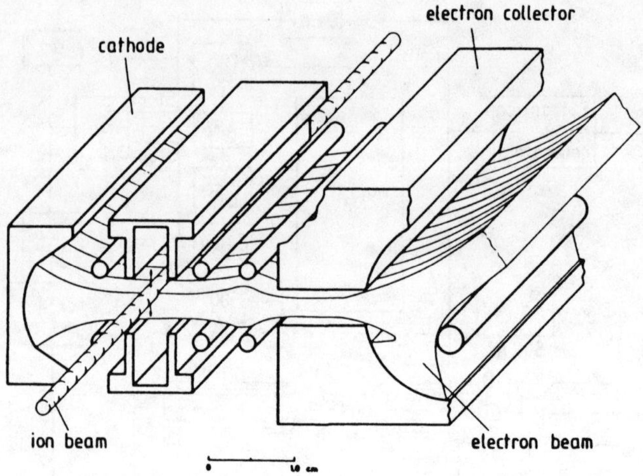

Fig. 1.4. A high-power, high-density electron beam for the crossed-beams experiments. The ribbon-shaped electron beam from a Pierce type gun crosses at its waist the target ion beam which stays at a fixed position and mechanically moves back and forth at a constant speed. From [1.12]

electron beams of 500 mA have been recently developed for the electron–ion collision experiments [1.12, 13] (Fig. 1.4). Such a high-intensity electron beam can provide much better signal-to-noise ratios and also, very fine structures in the cross sections, such as those due to Resonance-Excitation Auger Double-Emission (READE) processes (Sect. 2.1), can be observed.

1.2.2 Projectile-Ion Sources

There are a number of different ion sources [1.1, 2] or accelerators which can provide various ion beams from protons to uranium ions with a wide range of the ion energy ($< 100\,\mathrm{eV/u}$ up to a few GeV/u), namely, a simple high voltage power source, van de Graaff accelerator, cyclotron, linear accelerator, synchrotron and their combinations. In such combined apparatus, it is important to choose the electron stripping processes at a proper energy to maximize the acceleration efficiencies of the accelerator systems. Related information on the charge distributions of ions after passing through carbon foils can be found in the works [1.14, 15].

Low energy ion sources. With the recent developments of the ion-source technology, the ions with the energy less than 1 keV/amu, including bare ions such as Ar^{18+}, Kr^{36+} or even U^{92+}, can be extracted directly from the ion source of electron cyclotron resonance type, ECR [1.16] (Fig. 1.5a) or electron beam type, EBIS [1.17, 18] (Fig. 1.5b) or electron beam ion trap

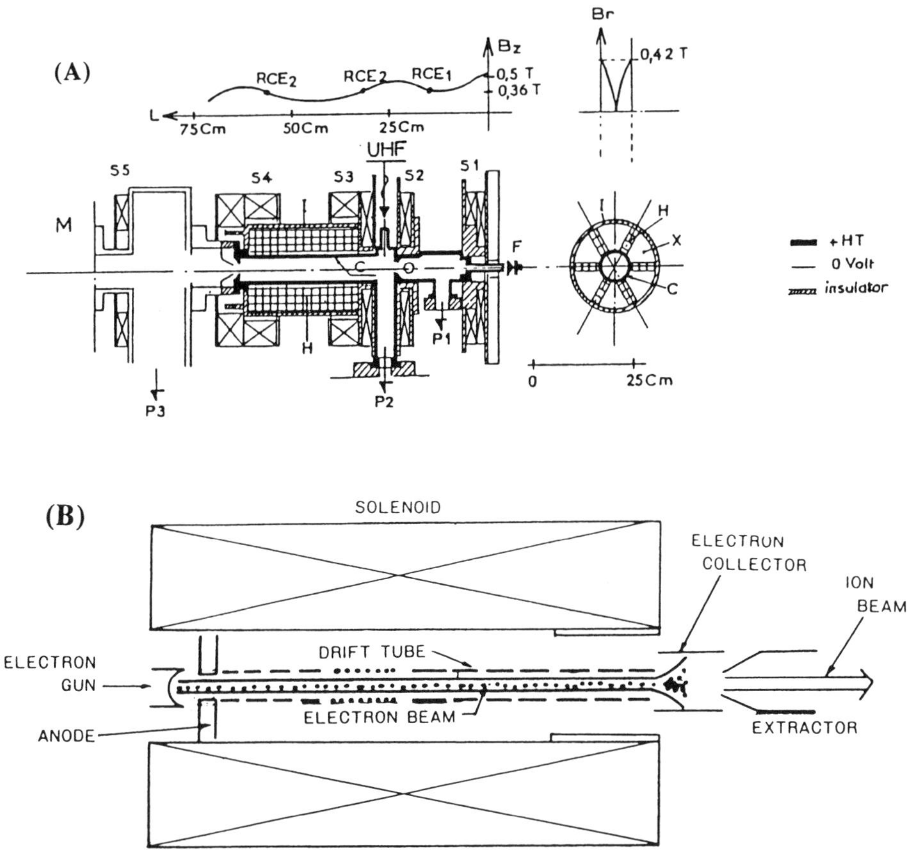

Fig. 1.5. Typical ion sources: (A) ECR ion source. The notations are as follows: F: iron shield, S_1–S_5: solenoids, P_1–P_3: pumping ports, E_1–E_2: extraction electrodes, UHF: micro-wave feed-through, C: multimode cavity, O: diaphragm, H: $SmCo_5$ hexapole magnets, I: insulators, B_z: solenoidal magnetic field along the Z-axis, X: X-ray shield, B_r: radial magnetic field. (B) EBIS ion source. The electron beam from a gun on the left is accelerated through an anode and passes through a series of drift tubes, meanwhile it is collimated with a strong magnetic field provided by a solenoid. The incident electron beam is diverged after the solenoid and dumped into the electron collector. The produced ions are extracted through an extraction field

EBIT [1.19, 20] (Fig. 1.6). The energy spread of these ions from the ion source which is sometimes critical in low energy collision experiments, such as the translational energy spectroscopy (Sect. 1.3.3), is much smaller than that from the accelerators. It should be noted that these ions can also be further cooled down to meV with various techniques and possibly in the

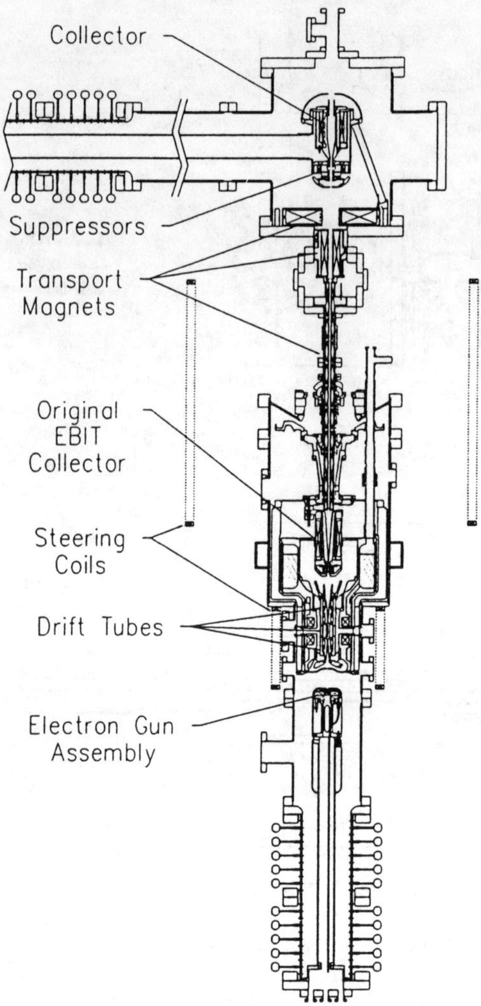

Fig. 1.6. The EBIT ion source. The primary electron beam formed at the electron gun assembly at the bottom of the figure is injected into the drift tubes. The produced ions are extracted through a collector, similar to that in EBIS

nearest future to neV-peV range and are suitable for the most accurate optical measurements [1.21].

Crossed- and merged-beams methods. Lower-energy collision experiments of the ions (1–10 eV) can be performed using the merged-beams technique where the projectile ion/electron and target ion move nearly parallel or completely parallel to each other [1.22–25]. Some detailed descriptions will be given in Sect. 1.4. Figure 1.7 shows a typical experimental setup for the crossed/merged-beams techniques used at low-energy collisions. The lowest energies achievable in this technique are limited by the energy spread of the original ion beams from the ion sources.

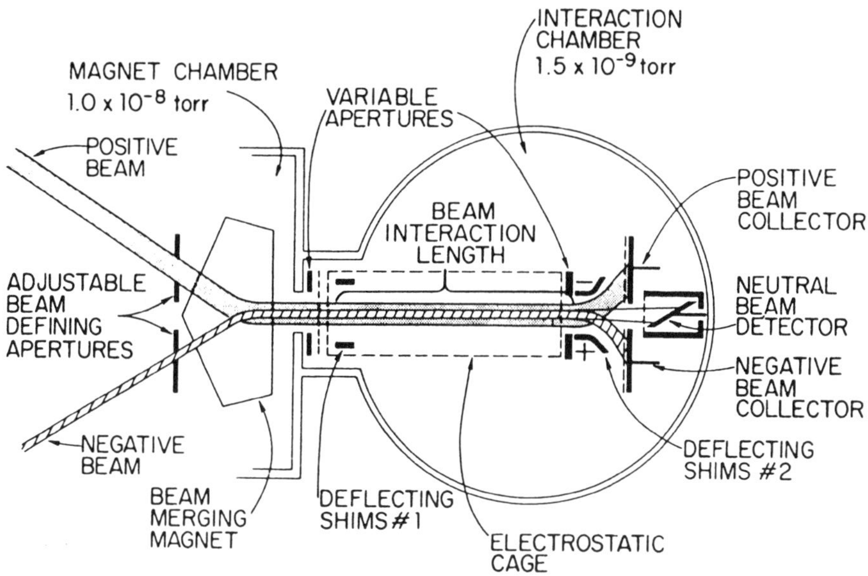

Fig. 1.7. Merged-beams techniques. The positive ions are introduced into the collision region after deflecting through a magnet, meanwhile the negative ions are also bent in the opposite direction through the same magnet and merged into the positive ions. After collisions, they are again separated with an electrostatic deflector. The collision products, neutral particles, are monitored with a neutral beam detector. From [1.26]

Octopole Penning Ion Gauge. Another technique has been developed to get very-low-energy ions down to eV range. First, the properly accelerated ion beam is sent into helium-gas atom targets where the incident ions lose their energy through the elastic collisions. During the collisions, the projectile ions have to be confined through special configurations such as an Octopole Penning Ion Gauge (OPIG) trap (Fig. 1.8) where the orbits of the confined ions are also shown [1.26].

Storage rings. Also it should be mentioned that recent development of the storage rings (Fig. 1.9) with the electron cooler systems result in small (longitudinal as well as transversal) energy spread of the projectile ions even at the high energy (GeV/u) range where very heavy, bare ions (up to U^{92+} ions) can be produced through electron stripping and, thus, high accuracies of X-ray observations become possible [1.27–35].

In the electron–ion collision experiments using highly charged, high-energy heavy ions at the ion storage rings with an electron cooler, very high resolution of the order of 10 meV can be performed in order to observe various structures of the cross sections due to resonance processes [1.36].

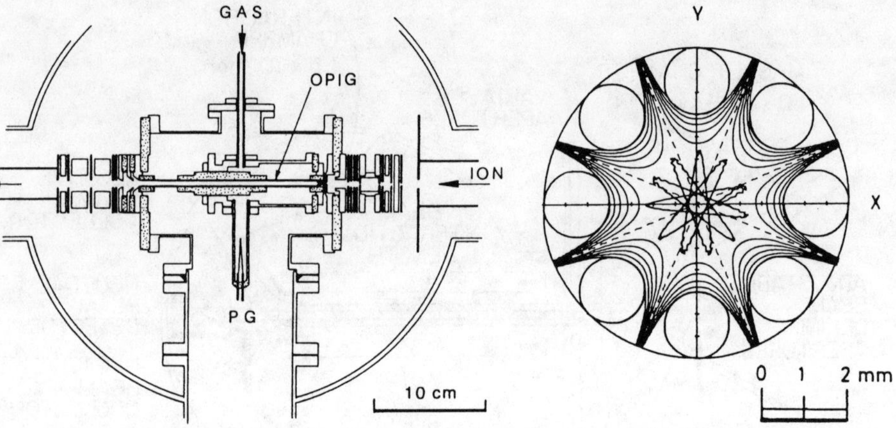

Fig. 1.8. (a) Collision apparatus with an OPIG ion trap. (b) A cut view of the OPIG showing the octopole electrodes, the electric field distributions and typical ion orbits. From [1.26]

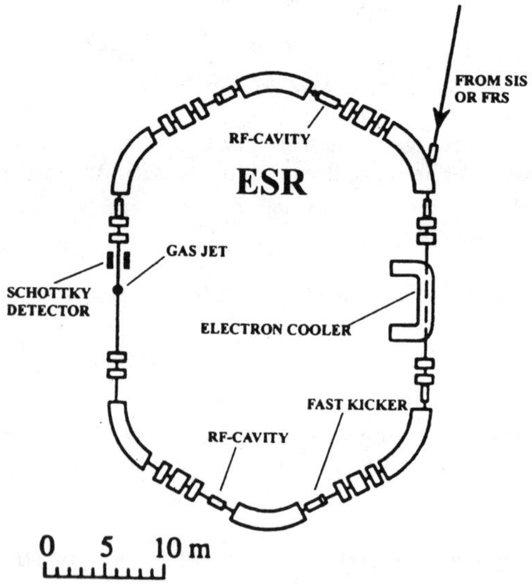

Fig. 1.9. An example of the ion storage ring, ESR, at GSI, Germany. From [1.27]

1.3 Detection and Measurements of Collision Products

A lot of the important information on collisions can be obtained through the observation of the energy and angular distributions of the collision products such the secondary electrons and photons as well as the energy gain/loss of the projectile ions. The coincidence techniques between these collision partners provide more detailed information on the collision dynamics.

1.3.1 Optical Measurements

Visible, ultraviolet (UV) and vacuum-ultraviolet (VUV) photons. The photon spectrometers are used to observe the photons emitted from atoms or ions in the excited states which are formed through collisions [1.37–39] (Fig. 1.10). In the visible region, the interference filters can be used to select the photons to be measured with a limited resolution. For better energy resolution measurements, the spectrometers such as the grazing-incidence type for the violet or ultraviolet region are necessary. The high-resolution spectroscopy allows us to find the detailed aspects (such as ℓ-distributions of the products) in the collision processes. The features of the photon spectroscopy strongly depend on the type of spectrometers. In particular, it is hard to keep those in the UV or VUV region in good and reliable shapes as the absolute sensitivities in the components such as gratings are often varied with the time and environmental conditions. Thus, it is necessary to take special care in obtaining the reliable photon emission cross sections.

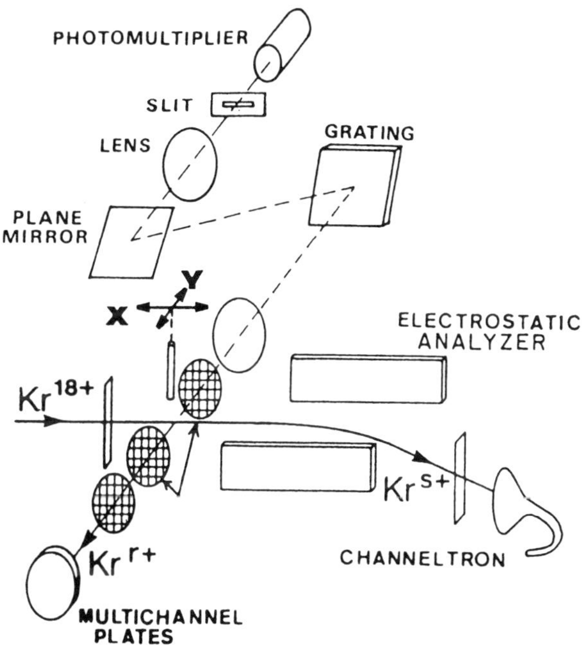

Fig. 1.10. A system for measuring the photons from Kr^{18+} ion-Kr atom collisions. The grating or crystal can be replaced with the interference filter for better efficiencies if high-energy resolution is not required. The coincidence measurements between the charge-selected projectile ions Kr^{s+}, the recoil ions Kr^{r+} and the photons provide the detailed information on the collision processes. From [1.38]

X-rays. X-rays can be detected with proportional counters for relatively low-energy photons below a few keV and with reasonable energy resolutions and energy discrimination against the backgrounds or by Si(Li) or ultrapure Si detectors with good energy resolution at keV region or by ultrapure Ge detectors with good energy resolutions at high energy up to 100–200 keV. Another type of compact solid-state x-ray detector based upon Cadmium-Zinc-Telluride (CZT) with similar resolutions and with much simple structures using Peltier coolers have been developed. When much higher resolution is necessary, the crystal spectrometers are commonly used though the overall efficiencies are low.

1.3.2 Secondary-Electron Measurements

The energy and angular distributions of the secondary electrons generated in collisions provide a lot of information on the collision mechanisms. To measure the absolute ionization cross section in projectile ion or electron impact, two procedures are generally needed:

i) total *apparent* cross sections $\Sigma m \sigma^i_m$ (σ^i_m is the partial cross section for production of m-times ionized ions) which can be determined through measurement of the secondary ions or electrons by a pair of parallel plate ion/electron collectors with an accuracy of 2–3% [1.40] (Fig. 1.11), and
ii) relative intensities or cross sections for ions with different charge through the time-of-flight techniques or magnet analysis [1.41]. The first technique, without knowing the energy distributions of the emitted electrons, has been well established and its accuracy of the absolute cross sections for apparent ionization is generally one of the best among various atomic measurement

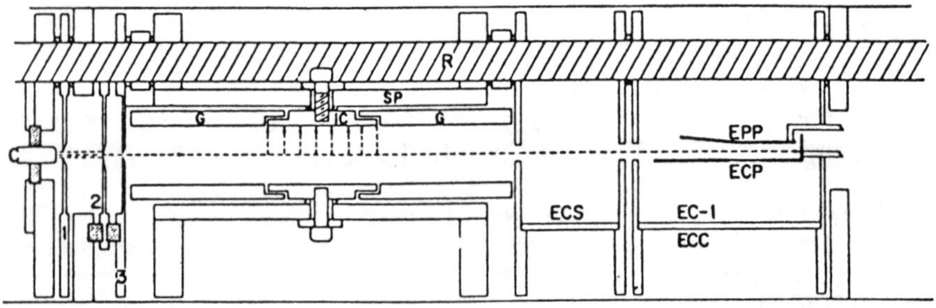

Fig. 1.11. A condenser plate technique to measure the secondary ions or secondary electrons produced in collisions. The uniformity of the electric field between a pair of the collectors (C) for collecting the secondary particles is ensured with the guard electrodes (G) on both sides of the collectors. This relatively simple system can provide the most reliable absolute (apparent) cross sections for ionization. From [1.41]

techniques, possibly 2–3%. Indeed, the cross sections measured in this way are used as the standard *absolute* ionization cross sections obtained in other measurements.

Secondary electrons. There are several types of the electron spectrometers [1.42]. Typical features of the simplest electron spectrometer consisting of a pair of the parallel plates is shown in Fig. 1.12. General structures of the energy spectrum of the secondary electrons produced in ion-atom collisions consists of two parts [1.43, 44]: i) the projectile-related part and ii) target-related part.

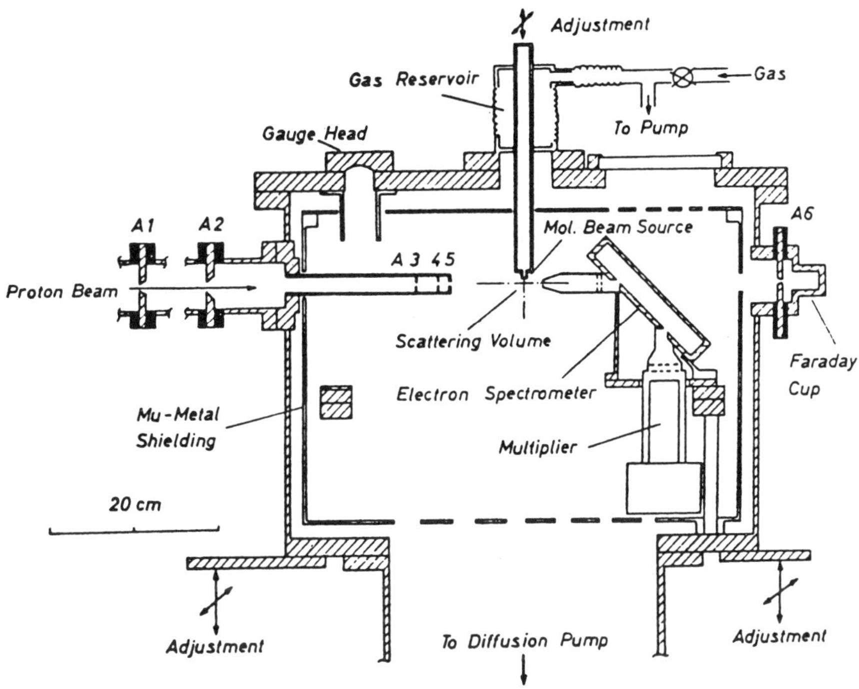

Fig. 1.12. A parallel plate electro static electron energy analyzer. The electrons enter into the analyzer at 45 degrees and after deflected are detected with a multiplier, meanwhile high energy projectile ions pass through a hole in the electrostatic analyzer. From [1.43]

As seen in Fig. 1.13, a large part of the secondary electrons have very low energy, namely, 1–10 eV, which originate from the target atoms through soft collisions with the projectile ions and are distributed nearly isotropically over all the angles. A small fraction of the secondary electrons have large kinetic energies which originate through the so-called close/hard binary encounter collisions with the projectile nucleus and thus the kinetic

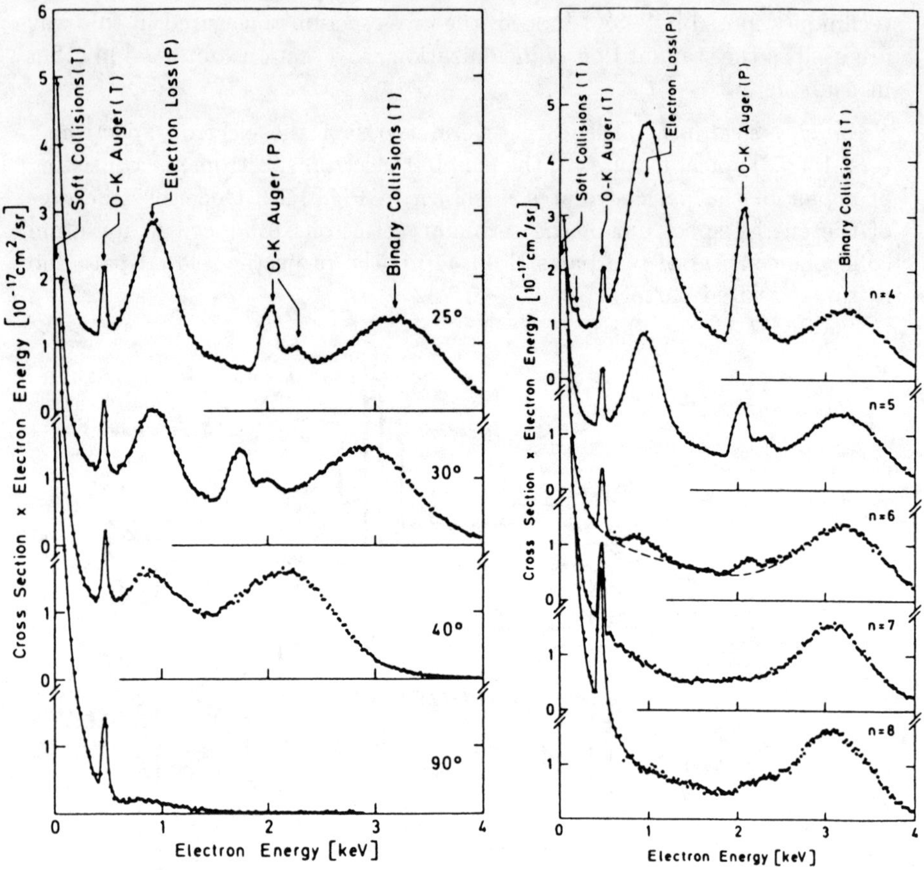

Fig. 1.13. Typical electron energy spectra produced in highly ionized projectiles colliding with gas targets. There are electrons originated from different processes: (i) the low-energy electron due to the soft collisions with roughly isotropic distributions, (ii) the highest-energy electron due to the hard (so-called binary encounter) collisions with the $\cos^2(\theta)$ energy distributions, (iii) the Auger peaks due to the targets with the isotropic distributions, (iv) the Auger electrons from the projectile ions energy-shifted and (v) the loss electrons from the projectile ions with the energy corresponding to the projectile ion velocity and scattered only in the forward directions. Here n is the principal quantum number of the resulting ion. From [1.44]

energy peaked at the position of $4[(1/2)m_e v_p^2 \cos^2(\theta)]$, where m_e, v_p and θ are the electron rest mass, the projectile velocity and the laboratory angle emitted, respectively. If the inner-shell electrons are ionized or excited, the Auger electrons are emitted from both projectile ion and target, which are generally isotropic over the angle relative to the projectile ion direction.

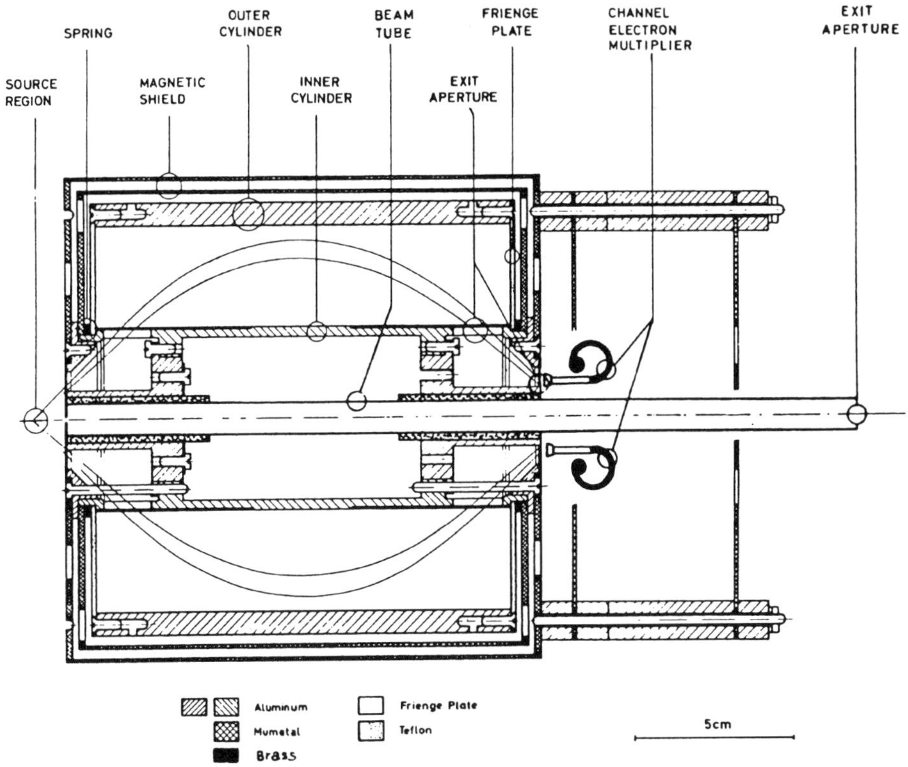

Fig. 1.14. A cylindrical mirror electron energy analyzer. The electrons are produced at the left end through collisions, enter into the mirror analyzer and finally are detected with a channeltron electron multiplier. On the other hand, the projectile ions, after collisions, pass through the center of the cylinders and exit through the aperture. From [1.45]

Another feature in the electron spectrum in heavy-projectile ion collisions is the electrons emitted at the forward angles at the energy corresponding to the projectile velocity, $(1/2)m_e v_p^2$, which is due to the electrons released from the projectile ions if they initially have some electrons or momentarily capture an electron into the continuum state of the projectile ions and, immediately after collisions, released from the projectile. When a further high-energy resolution is required, the retardation potential should be applied to such electron spectrometers or two spectrometers have to be used in tandem.

Instead of the parallel plate analyzer, the cylindrical-mirror analyzer with larger acceptance solid angles can be used with the advantage where the projectile ions are able to pass through the electron energy analyzer and thus can be performed some coincidence experiments with the charge-selected projectile ions [1.45,46] (Fig. 1.14).

Toroidal spectrometer. The toroidal electron spectrometer, combining with a position-sensitive detector, can detect the scattered or emitted electrons in a wide range of the scattered angles (roughly over 180 deg.) at a particular collision plane, shown in Fig. 1.15 [1.47]. This type of the toroidal spectrometer can be separated into two parts, namely, the first half is used for detecting the electrons and the second half for the recoil ions. Thus, two different particles, such as the secondary electrons and the recoil ions, can be detected simultaneously.

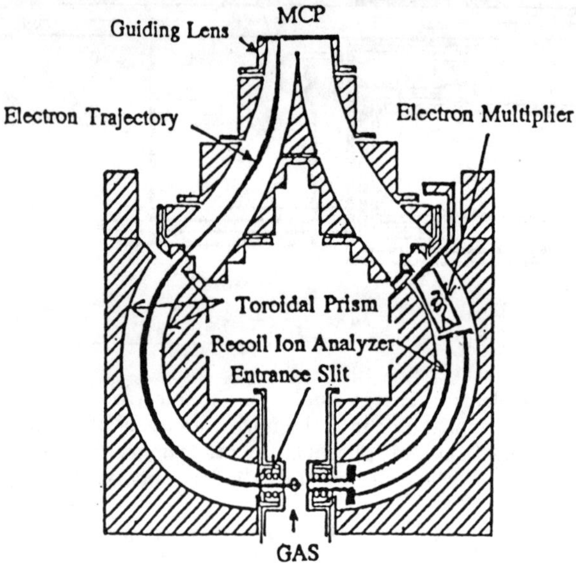

Fig. 1.15. A sheme of a toroidal analyzer. This special design tends to detect the scattered electrons and the recoil ions simultaneously over large scattering angles. From [1.47]

4π*-solid angle spectrometers.* Recently, a new type of the electron spectrometer with 4π solid-angle observation has been developed [1.48,49], without losing any information of the scattered angles and energy. This type of the analyzer can be used for simultaneously detecting both the recoil ions and the secondary electrons in coincidence (Fig. 1.16).

Zero-degree electron spectrometers. This technique [1.50,51], though similar to that described in Fig. 1.11, is quite unique and powerful, in particular suitable for the investigations of the secondary electrons emitted along the

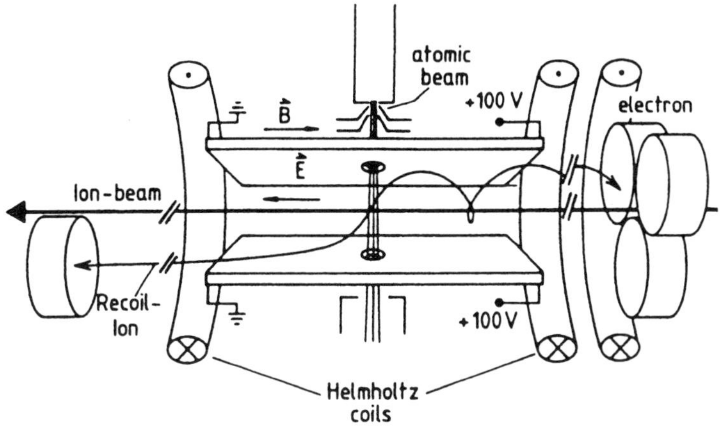

Fig. 1.16. A 4π solid-angle spectrometer for electrons and recoil ions. The secondary electrons move into the detector through the combined weak electric field and magnetic field distributed along the projectile ion direction. On the other hand, the secondary ions move into the other direction and finally reach the detector. From [1.48]

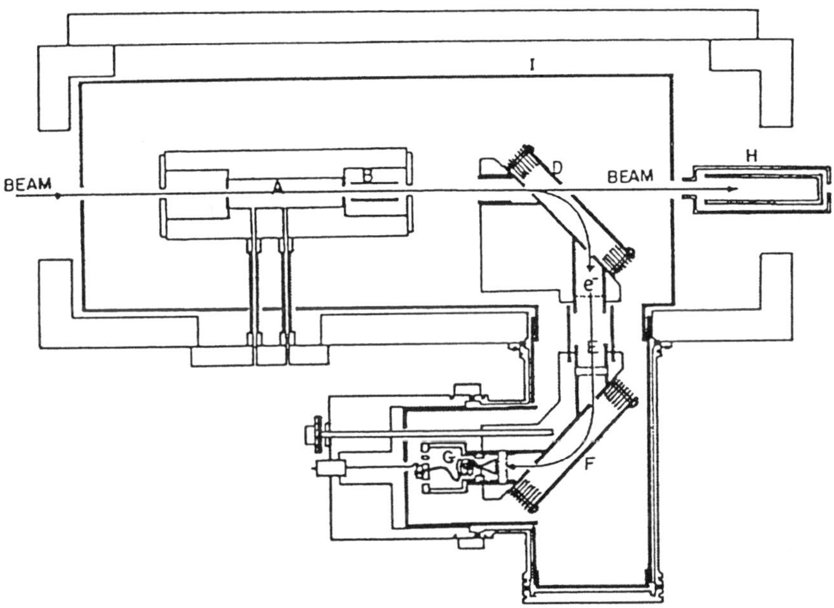

Fig. 1.17. A typical zero-degree electron energy spectrometer. The projectile ions come into a gas target A and, after slight tuning of their position with the deflector B, arrive at a Faraday cup H. The secondary electrons emitted from the projectile ions are energy-analyzed with a pair of the parallel electro-static deflectors D and F and detected with a channeltron G. I indicates the magnetic shield. From [1.52]

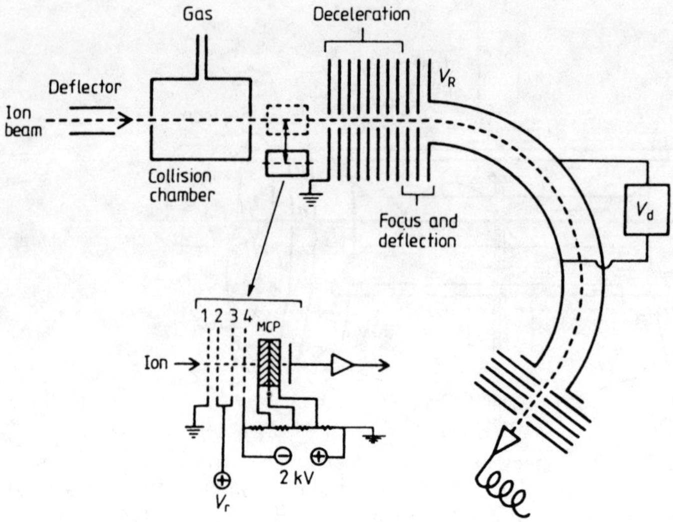

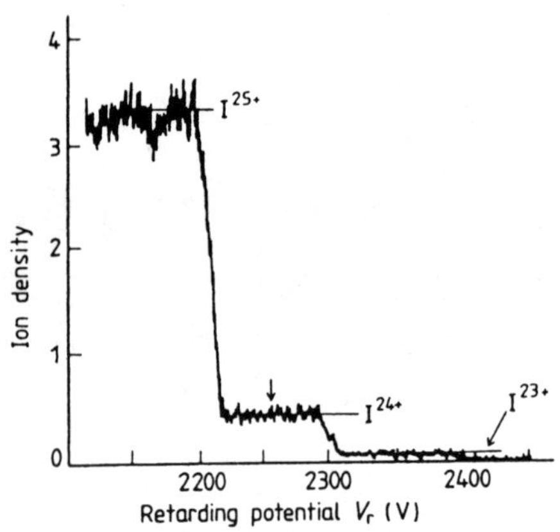

Fig. 1.18. Energy gain spectrometer with the retardation field which is shown in an enlarged inset. A typical variation of the intensity as a function of the retardation voltage is also shown. The steps in the intensity correspond to the charge of the incident ions. From [1.54]

beam incidence direction from the highly-charged projectile ions in the excited state formed in collisions such as ionization, excitation or electron capture-into-continuum processes (Fig. 1.17). Multielectron capture processes at low collision energies are known to result in the formation of highly excited Rydberg states which may decay via the autoionization or radiation processes or their combination. Very small energy of electrons emitted from the projectile ions (in the projectile frame) via the autoionization processes from high Rydberg states is transformed to high energy in the laboratory frame, which is much easier to measure with reasonable resolutions. The observed electron energy distributions and angular distributions provide important clues in understanding the collision mechanisms involving mul-

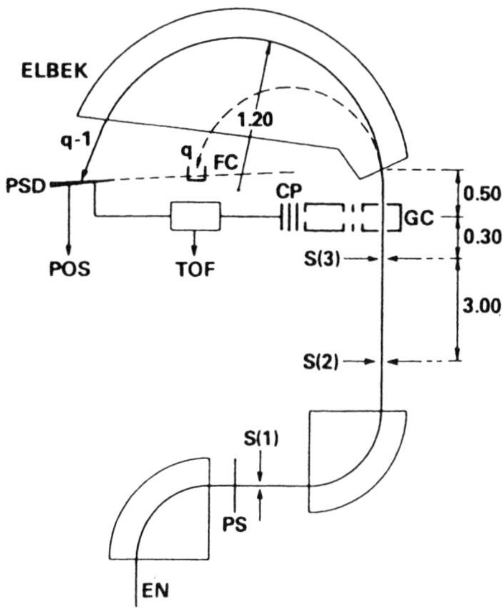

Fig. 1.19. A high energy-resolution magnet spectrometer capable of measuring the energy loss of 100 eV for 10 MeV projectile ions. High energy projectile ions with the charge q provided from an EN tandem accelerator pass through a number of the collimating slits S(1)–S(3) and collide with the target atoms in a gas chamber GC. After collisions, the projectile ions are energy- and charge-analyzed with a big ELBEK magnetic spectrometer and collected with a Faraday cup FC. On the other hand, the charge-changed projectile ions with the charge $(q-1)$ are detected with a position-sensitive detector PSD. The secondary ions produced in collisions are accelerated and detected with a channel plate detector CP. To identify the charge state of the secondary ions, a standard time-of-flight technique (TOF) is used. From [1.56]

tistep decaying processes. The detailed features of the zero-degree electron spectrometers and their results have recently been described [1.52, 53].

Ion Measurements. There are a lot of different techniques to identify the charge and the energy (change/loss/gain) of the projectile ions after collisions. There is the retardation technique for charge/energy measurements. This technique, though the energy resolutions are not good but is simple and convenient, has been used for a long time [1.54] and recently, successfully applied in low energy, highly-charged ion collision experiments to separate the projectile charge states after collisions (Fig. 1.18). Furthermore, this has recently been combined with time-of-flight (TOF) coincidence techniques of the secondary ions produced in collisions in order to get information on the correlation between the projectile charge and secondary ion charge [1.55].

Magnetic and electrostatic analyzer. This is one of the most standard techniques to separate the projectile charge in order to know the variations or fractions of the charge-changed projectile ions. Also, the magnet systems with large deflection angles resulting in high energy resolutions have been successfully used to measure a small change of the projectile energy down to 100 eV at the MeV projectile energies which enables us to know the energy losses in multiple-ionization processes as well as the inner-shell ionization [1.55–58] (Fig. 1.19).

Energy-selected translational energy spectroscopy. In order to observe a slight change in the projectile kinetic energy after collisions, high-resolution electrostatic energy analyzers, parallel plate, cylindrical, or spherical, are commonly used at low energy ($<$ keV/u) collisions, as in electron observation, by employing large deflection systems or retardation techniques; but the actual resolutions are often limited due to the original energy spread of the projectile ions produced in the ion source and barely sufficient to distinguish the difference of the n-distributions of the projectile ions in the electron capture processes. One of the examples is shown in Fig. 1.20 where a 45-degree parallel-plate electrostatic energy analyzer provides two parameters of the scattered projectile ions measured in the two-dimensional position-sensitive detector: scattering angle and energy. Typical energy resolutions of such systems are $\Delta E/E = 0.002$ and $\Delta\theta = 0.1°$. Such electrostatic energy analyzers are most widely used in low energy, highly-charged ion collisions [1.59]. This technique is also applied at high energies up to 200 keV region in order to in-

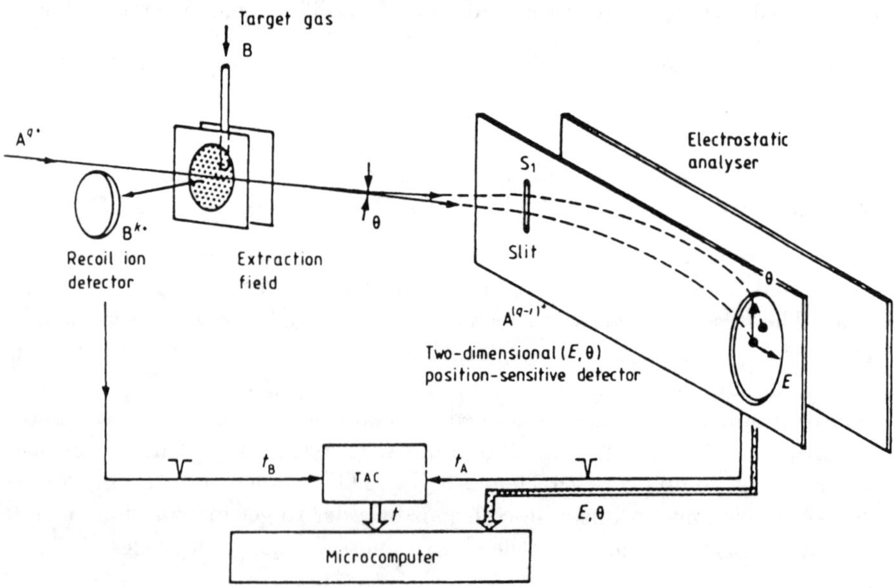

Fig. 1.20. A two-dimensional electro-static energy (E) and scattering angle (θ) analyzer system for projectile ions. The recoil ions are detected in coincidence with the E/θ-analyzed projectile ions. From [1.59]

vestigate the excitation and ionization through energy loss of the projectile ions. In this case, the energy loss spectroscopy is combined with the retardation technique by decelerating the projectile ions after collisions with the target gas atoms or molecules in order to get better energy resolution [1.60].

1.3.3 Recoiled Secondary-Ion Measurements

Most of the techniques that are practically similar to those mentioned above for high-energy collisions can be used to identify the charge distribution, the energy (momentum) distribution and the angular distribution of the recoil ions. Also, the time-of-flight (TOF) techniques are most commonly used at low-energy collisions. Because the energy of the recoil ions depends on where they are produced along the extraction direction due to the finite size of the incident beam, the flight time of the recoil ions to the detector from the collision region is varied. Yet a simple "space-focused" ion extraction system can improve such situations by properly choosing the extraction field strength, the acceleration field strength and the length of the free-drift region, independent of the position of the recoil ion produced along the ion extraction field due to the spatial spread of the projectile ions [1.61].

As the energy of the recoil ions provided during the collisions is very small (generally of the order of a few eV at best, though depending upon both the projectile ion and recoil ion charges as well as the impact parameters), special care has to be taken to obtain the accurate information of the recoil energies.

Recoil-ion energy measurements as micro-radian scattering measurements. In energetic collisions involving the outer-shell electron ionization, the projectile scattering is expected to be quite small (of the order of 10^{-6} radians). In measurements of differential scattering cross sections at such small angles, serious difficulties are encountered. The basic principle can be seen in Fig. 1.21 where the momentum conservation tells us that the transversal momentum component of the recoil ion, p_{rv}, perpendicular to the projectile ion direction, should be equal to the transversal momentum component lost from the scattered projectile ion, $p_{pv} = p_0\theta_p$ for a small θ_p where p_0 is the initial momentum of the projectile ion:

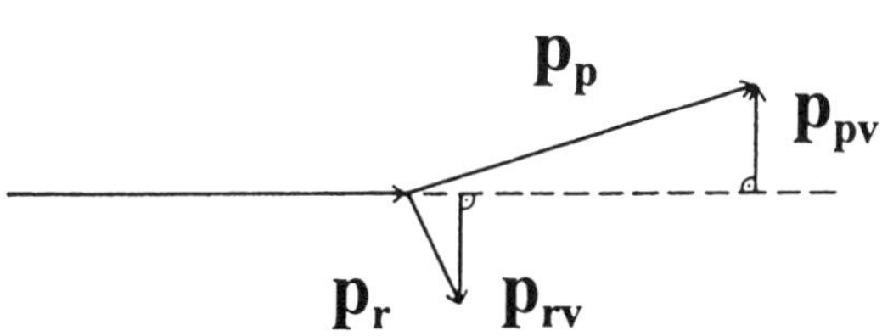

Fig. 1.21. Momentum diagram of energetic particle collisions: p_p and p_r are the momentum of the projectile ions and recoil ions, respectively, after collisions. p_{pv} and p_{rv} are their transversal momentum components. From [1.62]

$$p_{rv} = -p_{pv} = -p_0\theta_p \,. \tag{1.1}$$

Thus, determining the small scattering angles of the projectile ions can be reduced to measuring the transversal momentum of the recoil ions, p_{rv}, instead of the direct measurements of θ_p. This can further ease the requirements of some divergences of the incident projectile ions compared with that in the direct scattering angle measurements.

1.4 Crossed/Merged-Beams Techniques

When ions, instead of neutral atoms, are being used as target, their mass and charge state have to be specified after proper acceleration. Then the target ions are crossed by the projectile particles, either electrons or ions. By changing the crossing angle between the target ion and the projectile, relative velocity can be easily varied. As the densities of such ion targets are extremely low, specia techniques are requisite. Similar techniques can be also applied to neutral beams which are formed through electron capture into positive ions or electron detachment from negative ions.

1.4.1 Electron–Ion Crossing

In contrast to the thermal neutral atom targets which are easily available in gas forms, the ion targets have to be obtained through acceleration of proper ions and be crossed and excited or ionized by electron beam or another ion beam [1.63–65] (Fig. 1.22). Thus the effective densities of the accelerated ion targets are smaller, compared with that of the thermal neutral targets. The ion target densities at the keV energy are estimated to be equivalent to or less than roughly the order of 10^{-10} Torr. Therefore, the ultrahigh vacuum systems with 10^{-10}–10^{-11} Torr are requisite in the crossed beam collision experiments. Even in such low pressures, the signal-to-noise (S/N) ratios are usually quite low. To overcome low S/N ratios in the crossed-beams techniques, the time-modulation or chopping of at least one of them, but better of both beams, is necessary (Fig. 1.23).

Another important point in the crossed-beams techniques is to know the overlapping of both the projectile and target beams and to keep them constant during measurements [1.66] (Fig. 1.24). Since the first introduction, this technique has been developed in two typical forms: i) beam chopping and ii) one beam sweeping mechanically across the other beam, namely, animation [1.67,68] (Fig. 1.25).

Some of the processes, such as those involving the charge transfer between two colliding particles, can be investigated with much less trouble

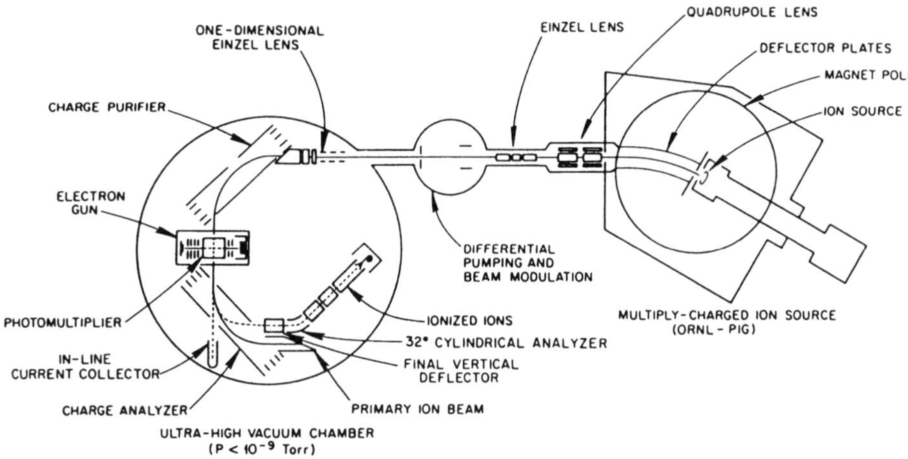

Fig. 1.22. A schematic setup of the electron–ion crossed-beams experiments. The ions produced in a Penning ion source are sent into an ultrahigh-vacuum collision chamber, charge-purified via a pair of the electrostatic plates and crossed through the electron beam. After collision the ions are again charge-analyzed. The primary ions are measured with a Faraday cup. The product ions are further charge-analyzed and, after refocusing, finally detected. From [1.65]

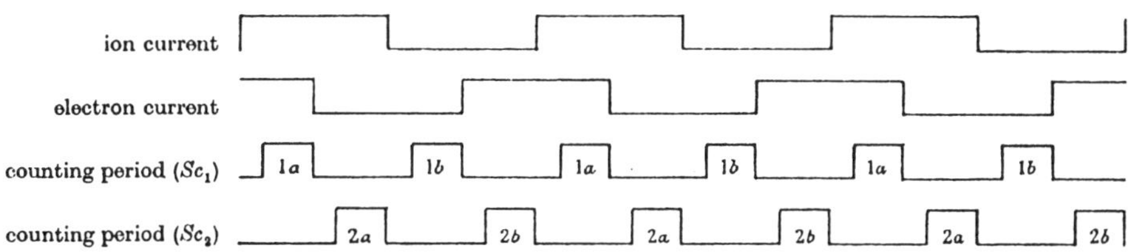

Fig. 1.23. Double beam-chopping scheme. The two beams are chopped with a quarter of the cycle shift. The signal counting is performed with the four different periods. The signals in the first counting period (1a), where the two beams are crossed, correspond to the sum of true signal (S_t) + background due to the electron beam (B_e) + background due to the ion (B_i) + background without both beams (B). The signals in the second (2a), where the ion beam is on but the electron beam is off, correspond to (B_i + B). The signals at the third (1b), where both beams are off, correspond to (B) and the signals in the fourth, where only the electron beam is on, correspond to (B_e + B). Therefore, the subtraction of the signals of the sum (2a+2b) from the sums (1a+1b) gives the true signal S_t. In the actual situations the pulsing time sequence of one of the beams is periodically shifted by a half cycle in order to avoid the beam fluctuations during a long period of the counting time. From [1.64]

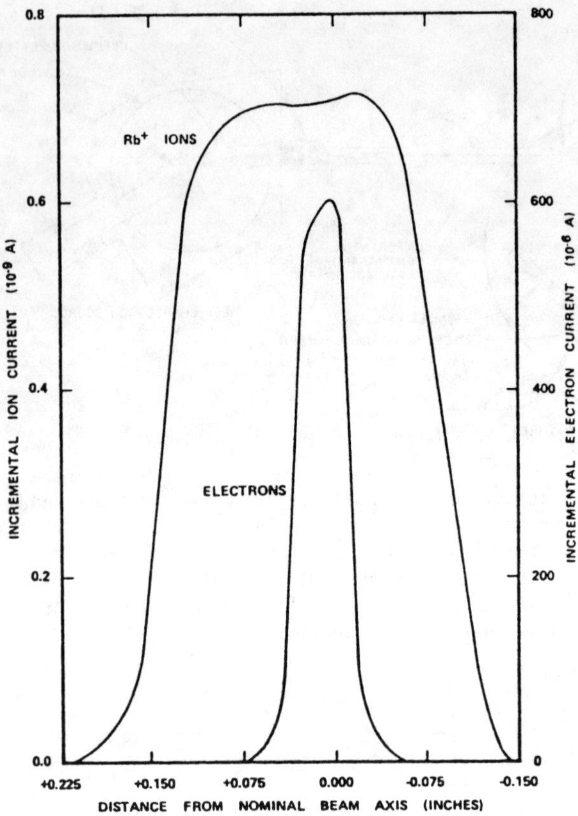

Fig. 1.24. A typical example of the beam-overlapping in electron–ion collisions. From [1.66]

through the coincidence between both the product beams in the crossed-beams experiments.

A serious problem in the crossed-beams measurements is related to the fact that the target beam often includes not only the ground-state particles but, in most cases, a significant fraction of the metastable state particles [1.69,70]. Generally, as the cross sections for the metastable state beams are far larger (often more than one order of magnitude) than those for the ground-state beam, a very small fraction of the metastable beams contained in the targets significantly influence the observed cross sections. Of course, the fraction of such beams can be somehow controlled and changed by mixing some different gases into the ion source gas. Their fractions can be estimated reasonably well (i) as far as the cross sections for both the ground and excited state ions are already known and (ii) by measuring the attenuation of the target ions as a function of the second-collision gases.

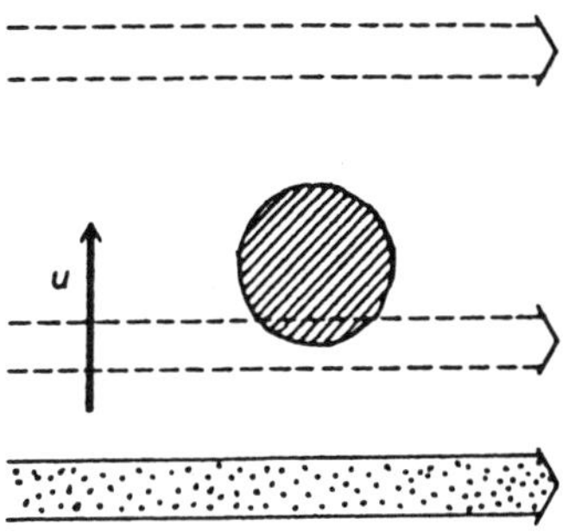

Fig. 1.25. The beam animation technique. The target beam at a fixed position indicated with a shaded circle is animated by the projectile beam which moves up and down at the constant velocity as indicated with an arrow. From [1.68, 69]

1.4.2 Electron-Neutral Crossing

The crossed-beams techniques are applied not only to the charged particles but also to the fast atomic particles which cannot be easily obtainable in a vapor form but are neutralized after acceleration of the proper ions. In particular, this technique is most preferable to obtain the reliable cross sections for the metallic targets which are not easy to get proper pressures of vapors under the well-controlled, stabilized conditions. For example, the metastable hydrogen beam, H(2s), can be obtained up to 20% of the total neutral hydrogen atoms formed by passing protons through some vapors, favorably alkalis such as Cs. By applying the quenching field, the contribution of H(2s) and other excited hydrogen atoms can be separated from that of the ground state H(1s) hydrogen beam [1.71, 72]. More recently, these neutralized species are used in the crossed-beams measurements of electron impact excitation cross sections for the excited Ar atoms [1.73]. As mentioned before, to get lower collision energy, the merged-beams techniques are usually used by putting both the projectile and target-ion beams nearly parallel. In this case, overlapping two beams over the long collision region is much more difficult.

1.4.3 Ion–Ion Crossing

Ion–ion collisions seem to be technically quite similar to electron–ion collisions [1.74, 75]. Yet due to the limitation of the ion current densities, the experiments are far more difficult to separate signals from backgrounds due to collisions with residual gases. Indeed the signal/noise ratios are usually 1/100 or less. To reduce such backgrounds and the scattered ion particles from the edges and surfaces, mostly originating from very weak tail distributions of the incident ion beams, both the product ions, which are estimated to be of the order of 10^{-8} or less of the primary ion intensities, have to be deflected away from the plane of the primary projectile ions and energy-

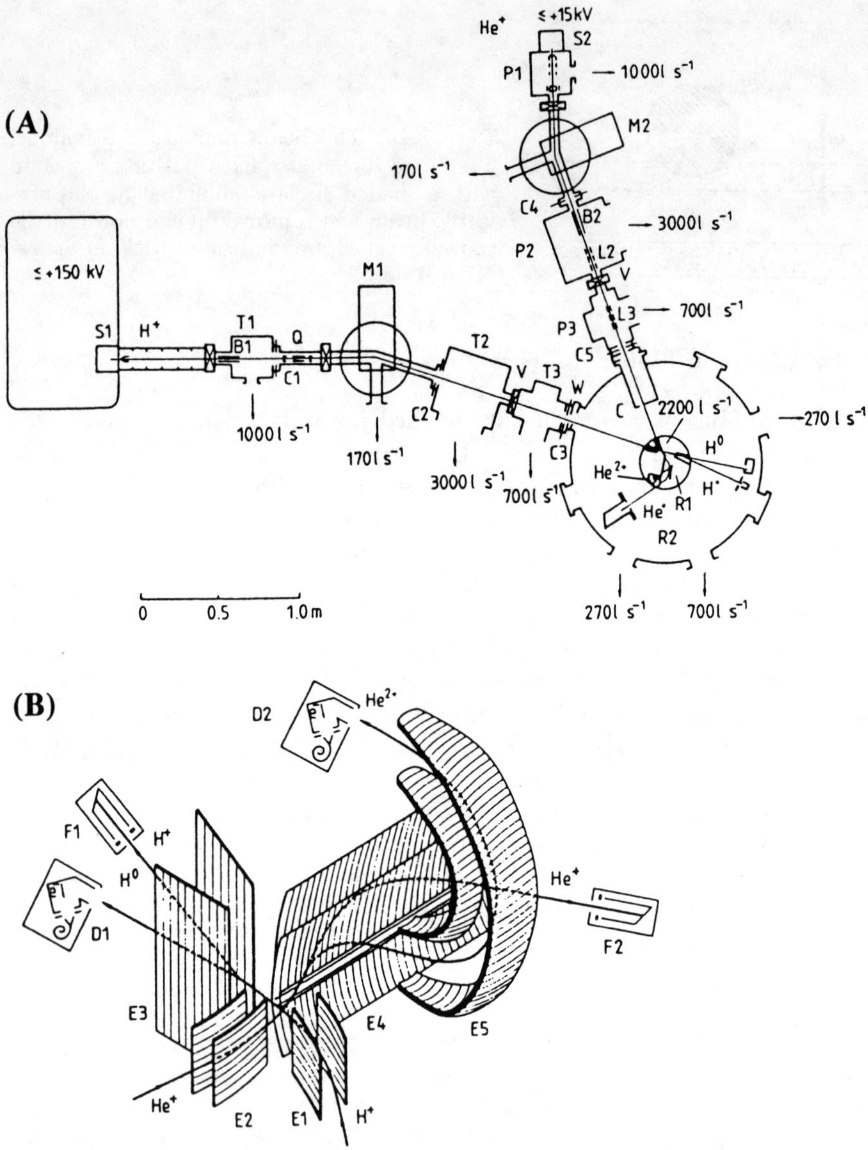

Fig. 1.26. (A) An overall view of ion–ion collision experiments. The target He^+ ions are extracted at 15 keV and are crossed with the projectile H^+ ions accelerated up to 150 keV. (B) A schematically expanded collision analyzer system. The collision product H^0 produced via the electron capture is charge-separated and detected with D1. On the other hand, the product ion He^{2+} is first charged-separated through the first spherical electrostatic analyzer and then energy-analyzed through the second electrostatic spherical analyzer. From [1.75]

selected, namely, twice the deflections; the first for the charge separation and the second the energy selection after collisions (Fig. 1.26).

1.5 Coincidence Techniques

In the present-day experiments, the time-of-flight techniques are most commonly used to identify the recoil particles including their energy and charge and possibly the energy-changed or scattered-projectile ions. The coincidence between at least two products, for example, charge-selected projectile ion-secondary ion, and multiple coincidence (for example, charge-selected primary ion–secondary ion–electron) provide more information on the collision dynamics both in low and high energy collisions.

1.5.1 Coincidence between Recoil Ions and Electrons emitted from Projectile Ions

The recoil ions produced in collisions are extracted and accelerated and charge-analyzed, meanwhile the secondary electrons emitted from the projectile ions after forming the multiply excited states are energy-analyzed through a cylindrical spectrometer (Fig. 1.27). The coincidence between

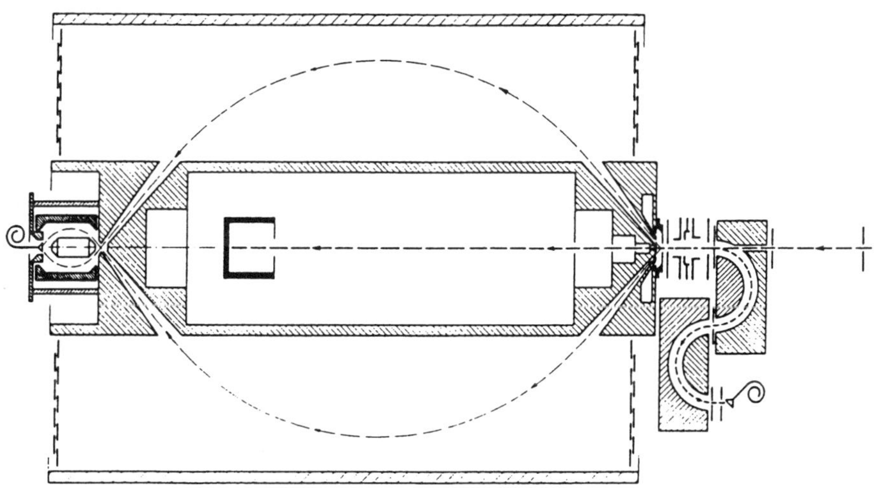

Fig. 1.27. A scheme of the cylindrical mirror electron energy analyzer at the center and the double hemispherical ion spectrometer at right. The incident projectile ions enter into the collision chamber from right and the collisions with the target gas occur at the right end region. From [1.46]

the charged-selected recoil ion and the emitted Auger electrons gives information on the multielectron capture processes, the number of the captured electrons and the configurations of the projectile ion after electron capture. Figure 1.28 shows a typical coincidence spectrum in 70 keV N^{7+} + Ar collisions where the largest contribution is due to the double-electron capture forming mostly the doubly excited state ($n_1, n_2 = 4, 4$) where n_1 and n_2 indicate the principal quantum numbers of two captured electrons and three electron capture, followed by a single-Auger electron emission forming the $(3, n)$ states and four electron capture followed by double Auger electron emission forming the $(3, n)$ states [1.77].

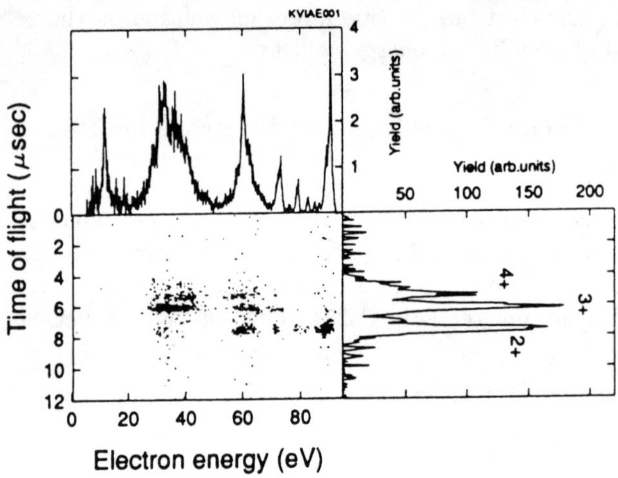

Fig. 1.28. A typical coincidence spectrum between the recoil ion and the secondary electrons in 70 keV N^{7+} + Ar collisions. The upper figure shows the single spectrum of the secondary electron emitted from the doubly excited states $(4, 5)$, $(3, n)$ and $(4, 4)$ and the spectrum at the right end shows the recoil ion charge distribution. The coincidence spectrum shows that the two electrons are captured dominantly into the $(4, 4)$ states. From [1.77]

1.5.2 Coincidence between Charge-Changed Projectile Ions and Photons

Doubly excited states, which are formed through the Resonant Transfer Excitation (RTE) into the projectile ion from a quasi-free electron in target atom, accompanied simultaneously with the excitation of an inner shell electron in the projectile ion, can decay either (single- as well as double-) radiative (RTER) or nonradiative Auger (RTEA) emission processes [1.78–82]. In principle, the RTE is a process similar to the dielectronic

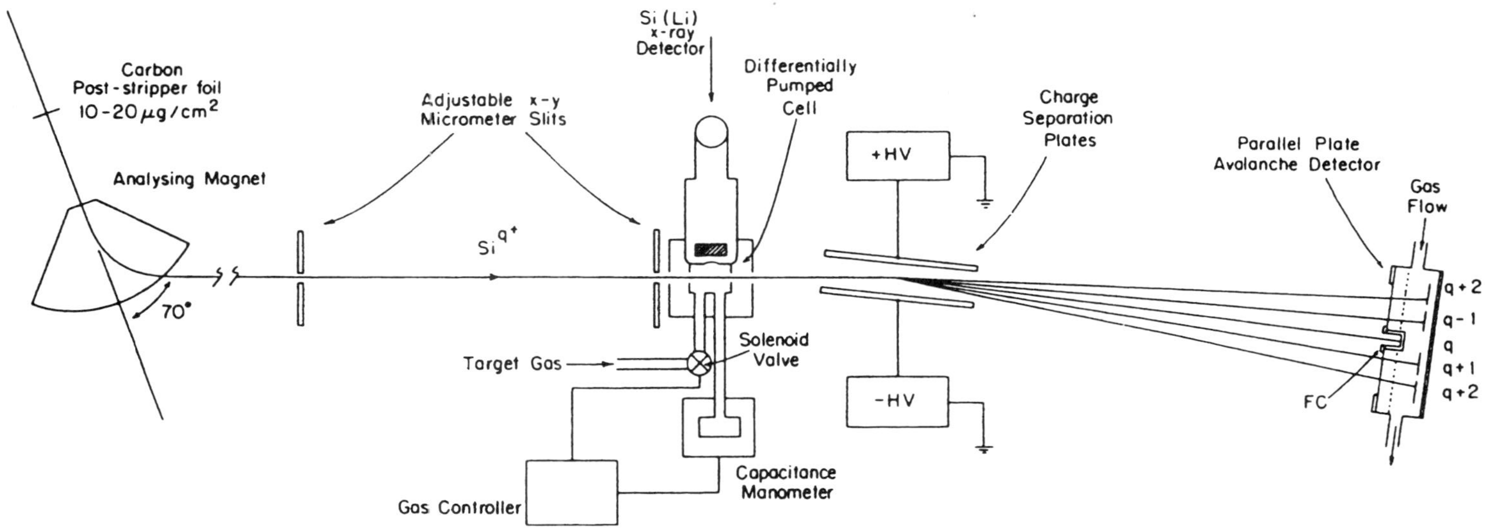

Fig. 1.29. An experimental setup for RTE measurements. The x-rays emitted from the projectile ions are observed with a Si(Li) detector at the gas target region, meanwhile, the charge-selected projectile ions are detected with an avalanche detector at the right end. From [1.79]

recombination involving a bound (quasi-free) electron in the target with the Compton energy distribution profile, instead of a free electron. Experiments determining RTE are based upon the coincidence between x-rays from the projectile and charge-changed projectile ions (Fig. 1.29). On the other hand, no resonance has been observed (nonresonant excitation, NTE) for light-projectile-ion impact such as $F^{8+}(1s)$. In this process, the doubly excited states are formed through projectile-electron excitation coupled with the target-to-projectile electron transfer to an excited state during a single collision [1.83].

1.5.3 Coincidence between Charge-Changed Projectile Ion and Recoil Ion

This method is often called the *Q-spectroscopy.* For example, in a single electron capture process, such as

$$\mathrm{He}^{2+} + \mathrm{He} \to \mathrm{He}^{+}(n\ell) + \mathrm{He}^{+}(n'\ell') + Q\,, \tag{1.2}$$

two heavy particles play an important role. Here, Q represents the energy change (defect) before and after collision. Thus, the full knowledge of their three momenta, p_x, p_y, p_z, before and after collisions assuming the target atoms at rest before collision, provides information in determining the complete momentum balance and thus the whole collision dynamics. In small angle scattering processes, such as in the ionization or electron capture, the momentum and energy conservation law for the two particles before and after collisions give the following relationship on the longitudinal $p_{r\|}$ and transverse p_{rv} momenta of the recoil ion after collision [1.84–87]:

$$p_{r\|} = -Q/v_0 - m_e v_0/2\,, \tag{1.3}$$

$$p_{rv} = -p_{pv} = -p_0 \sin(\theta_p)\,, \tag{1.4}$$

where Q is the so-called Q-value of the reaction, m_e the electron rest mass, v_0, p_0 and p_p the initial velocity, the initial momentum and the final momentum of the projectile ion, respectively, and θ_p the scattered angle of the projectile ion which is related with the transferred momenta as

$$\tan\theta_p = p_{pv}/p_{p\|} = p_{pv}/p_0 = -p_{rv}/p_0\,. \tag{1.5}$$

The second term in (1.3) is due to the electron mass transfer from the target to the projectile ion (obviously, if j-electrons are transferred, this term should be multiplied by j). Thus by measuring $p_{r\|}$ and p_{rv} with the known initial parameters, Q-values can be obtained. Therefore, the final captured and excited electronic states in the projectile ion become known. A typical experimental setup is shown in Fig. 1.30. Information on the projectile ion (charge state, scattering angle) can be registered on the two-dimensional

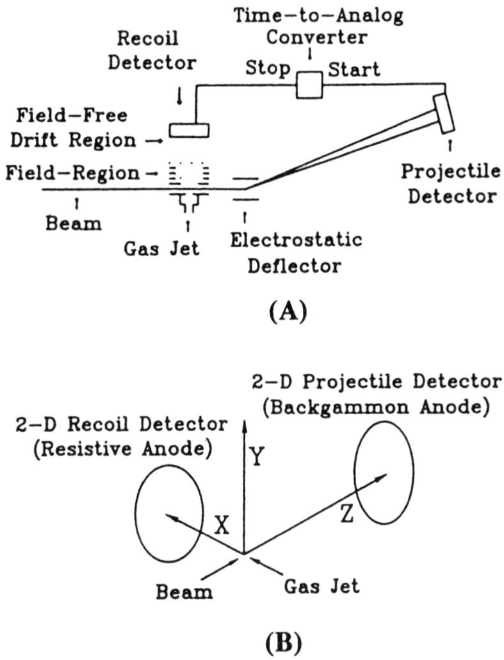

Fig. 1.30. (A) Time-of-flight detection system for the charge-selected projectile ion-recoil ion coincidence. (B) The coordinates of the projectile and recoil detectors. From [1.85]

position-sensitive projectiie detector after passing through the electrostatic deflector. Meanwhile, information on the recoil ions (charge, momenta) can be collected on the second two-dimensional position-sensitive recoil detector after their extraction through a proper uniform electric field, in combination with a time-of-flight coincidence with the scattered projectile ions. $p_{r\parallel}$ is determined by measuring the longitudinal displacement of the recoil ions along the z-axis (namely, the incident direction of the projectile ion). On the other hand, in order to determine p_{rv}, it is necessary to measure two momentum components, namely, p_{rx} and p_{ry} components along the x- and y-directions, respectively. p_{ry} can be obtained through the displacement of the recoil ion and time-of-flight in conjunction with the projectile ion, meanwhile, p_{rx} is obtained through TOF with the projectile ion. Finally, $\sqrt{p_{rx}^2 + p_{ry}^2}$ gives p_{rv}.

The momentum resolution in such a system is limited mostly by the thermal motion of the target jet in the z- and y-directions (and also by the projectile ion beam size in the y-direction) and the Maxwellian-Boltzmann distribution of the target flow at the intersection of the projectile and the jet in the x-direction. New cold targets have been recently developed to overcome this problem, and further can reduce the effects of this thermal energy spread (Sect. 1.1.3).

1.5.4 Multiple Coincidence among Secondary Ions, Secondary Electrons and Projectile Ions

In order to do a complete experiment involving the emission of electrons such as the ionization processes, it is necessary to measure also the momentum distribution of the secondary electrons emitted in collisions. In combining the cold gas target mentioned above, a new recoil-momentum spectrometer has recently been developed, as shown in Fig. 1.31 [1.88–91]. Here the projectile ions come into the collision jet region where, through a weak longitudinal electric field generated on a pair of ceramic plates coated with resistive layers, the secondary recoil ions produced are accelerated along the direction of the projectile ion with a large Larmor radius under the longitudinal magnetic field provided by a pair of Helmholz coils, and finally reach the two-dimensional position-sensitive ion detector with almost 100% of the solids angles covered. On the other hand, the secondary electrons are accelerated into the opposite direction with a much smaller Larmor radius and finally arrive at three electron detectors, also two-dimensional position-sensitive electron detectors, which cover roughly 50% of the solid angles (see similar spectrometer in Fig. 1.31. Using this detection system, the recoil ion and secondary electron momenta as well as the change of the projectile longitudinal momentum in He^{2+} ion production under 3.6 MeV/u Se^{28+} + He collisions have been determined, as shown in Fig. 1.32.

Using this technique, the small transversal as well as longitudinal momentum transfer in multiple ionization processes and also the state-selective scattering of the projectile ions involving the electron capture can be determined [1.90].

A similar system has been extended to determine, through triple coincidence technique (though no cold gas target is used), all the momenta involving three particles (charge-selected projectile ion, recoil ion and secondary electron) in collisions based on three two-dimensional position-sensitive detectors, as shown in Fig. 1.33. Here the uniform magnetic field for efficiently and accurately collecting the slow electron is provided with three pairs of the Helmholz coils. To perform such multiple coincidence measurements, it is a prerequisite to develop the efficient data-taking systems as well as data-processing systems based upon fast computers which are not discussed here.

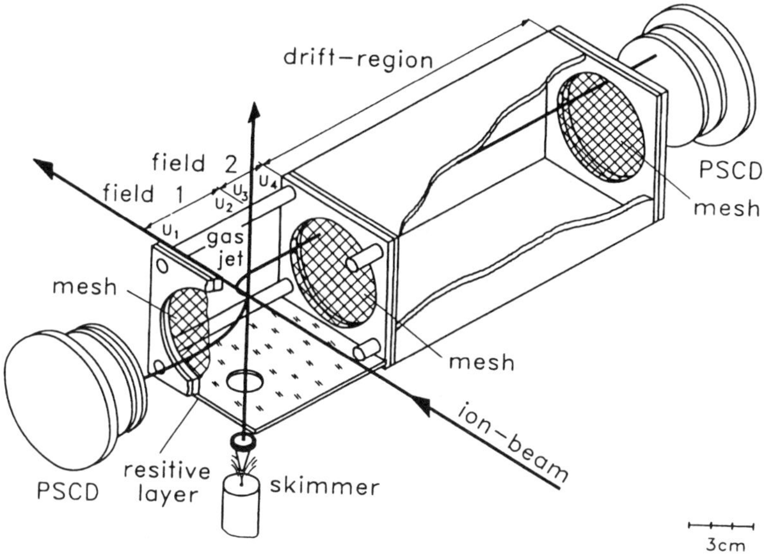

Fig. 1.31. A scheme of the recoil-momentum spectrometer for both the recoil ions and the secondary electrons. Here the recoil ions and the secondary electrons, produced in collisions of energetic projectile ions through a cold-gas jet, are accelerated into the opposite directions through a weak electric field generated along the semiconducting resistive layer on the ceramic plates and under the magnetic field of a pair of the Helmholz coils and finally arrived at the position-sentive detectors PSCD at both ends. From [1.88]

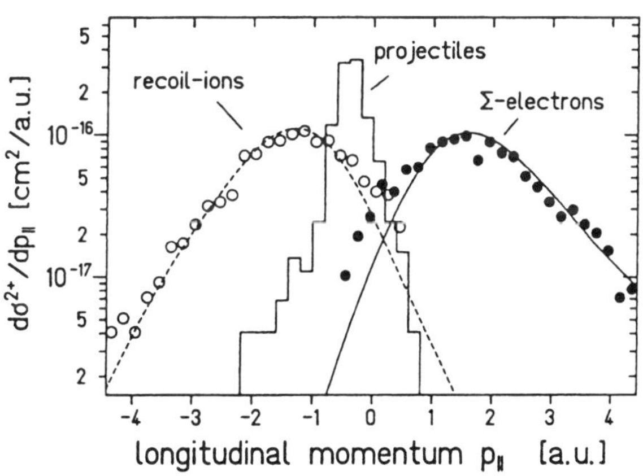

Fig. 1.32. The observed momentum distributions of the recoil He^{2+} ion and electrons produced in 3.6 MeV/u Se^{28+} ion colliding with a He target. From [1.89]

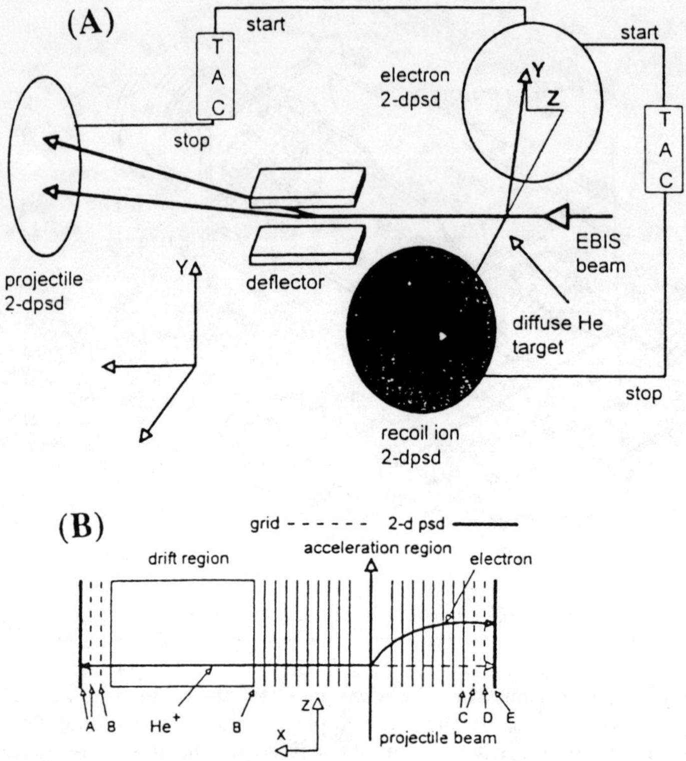

Fig. 1.33. A scheme of the triple-coincidence technique with three two-dimensional position-sensitive detectors for the projectile ions, recoil ions and secondary electrons (A). (B) shows the field as well as the path of the secondary ions (He^+) and electrons. The projectile ions pass through the collision region from the bottom (B). A and E in (B) are two-dimensional position-sensitive detectors for the secondary ions and electrons, respectively. From [1.92]

2. Multielectron Processes Induced by Charged Particles

Multielectron processes arising in electron–atom and ion–atom collisions are of great interest from both experimentalist's and theoretician's viewpoints. Among them are multiple excitation and ionization, multielectron capture by highly charged ions, ionization with simultaneous excitation of the target electrons and others. The present status of theory and experiment related to these processes will be considered in the following chapter.

2.1 Multielectron Ionization by Electron Impact

Multiple Ionization (MI) of atoms and ions by electron impact is an important mechanism playing a key role in various physical applications such as plasma kinetics, development of new laser schemes, modeling of laboratory and astrophysical plasmas, charge-state evaluation of atoms exposed to an electron beam and others. Several review articles and books [2.1–10] are devoted to experimental and theoretical studies of multielectron processes induced by electron impact, including multiple ionization.

At present, cross sections σ_m of m-fold ionization of atoms and ions by electrons

$$e^- + \mathrm{A} \rightarrow e^- + \mathrm{A}^{\mathrm{m}+} + me^-, \quad m \geq 2 \tag{2.1}$$

have been measured for most of neutral atoms and a lot of species, negative and positive ions, including highly charged ions (see, e.g., [2.4, 11–14] and references therein). The results cover the incident electron energy range from the ionization threshold up to about 10-20 keV.

The most accurate measurements have been performed using the crossed-beams technique (Chap. 1) and achieved a high degree of confidence. The absolute total uncertainties of the order of a few percent are frequently obtained even in the smallest investigated cross section range $\sigma_m < 10^{-22}$ cm^2. Although, in some cases, the available experimental data on σ_m are inconsistent, the experimental material allows one to make general assumptions concerning the behavior of the cross sections and their dependence on the threshold ionization energy, the number of ejected electrons and the electron-impact energy. In contrary to the large volume of the experimental data obtained, a theory of multiple ionization of atoms and ions by electron impact still suffers from the lack of the quantum-mechanical treatment even for two-electron ionization. This is mainly connected with the fact that the independent electron model, used in ion–atom collisions, fails in the description of multielectron transitions induced by electron impact. Therefore, to predict MI cross section behavior and their values, semiempirical formulas are often used (Sect. 2.1.2).

2.1.1 Main Features of MI Cross Sections

The measured threshold energies E_{th} for m-fold ionization cross sections coincide within 20% with the energy I_m required to remove m outermost electrons from the target A^{q+}:

$$E_{\text{th}} \approx I_m = \sum_{q'=q}^{q+m-1} I_{q',q'+1}, \tag{2.2}$$

where $I_{q',q'+1}$ is the one-electron ionization energy from the charge q' to $q'+1$. For single ionizaton, I_1 is the first ionization potential; the minimal energy I_2 for double ionization is the sum of the first and the second ionization potentials of the target, and so on. For example, the minimal energy I_3 required to ionize three electrons in Ar atom is estimated to be: $I_3 = I(\text{Ar}) + I(\text{Ar}^+) + I(\text{Ar}^{2+}) = 15.8\ \text{eV} + 27.6\ \text{eV} + 40.9\ \text{eV} = 84.3\ \text{eV}$. Minimal ionization energies I_m can be estimated using the tables for $I_{q,q+1}$ for atoms and ions given in [2.15, 16]. The minimal energy I_m to ionize all N electrons from a neutral atom is well described by the statistical Thomas-Fermi formula

$$I(m = N) = 16 N^{7/3}\,\text{eV}. \tag{2.3}$$

A typical example of m-electron ionization cross sections is given in Fig. 2.1 for e + Cu collisions. Partial cross sections σ_m decrease relatively slowly with the incident electron energy but strongly fall off with the number m of ejected electrons. As a rule, the main contribution to the total ionization cross section comes from single and double-ionization processes. In general, the multiple ionization by electron impact takes place via different processes:

(*i*) *direct* processes: simultaneous ionization of m outermost electrons (*Direct Multiple Ionization*, DMI):

$$\text{e}^- + \text{A}^{q+} \to \text{e}^- + \text{A}^{(q+m)+} + m\text{e}^-, \tag{2.4}$$

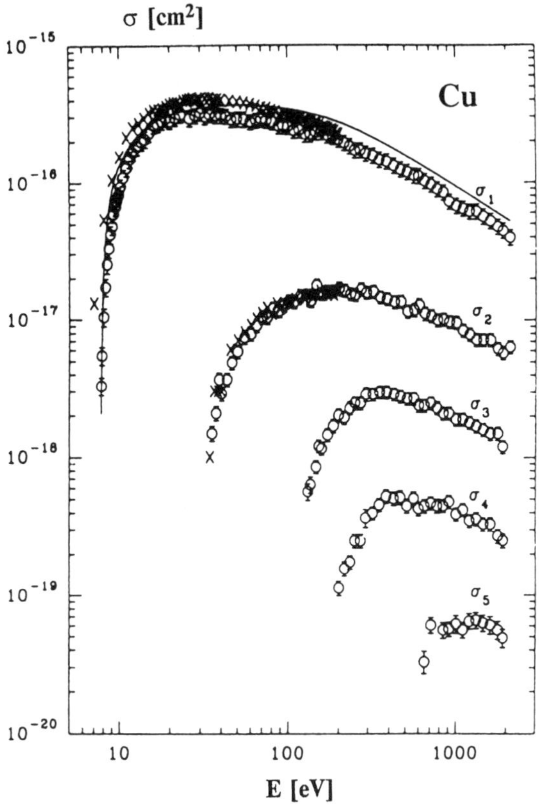

Fig. 2.1. The m-fold ionization cross sections σ_m of Cu by electron impact. Experiment: open circles – [2.17], crosses – [2.18]. Theory: solid curve – single-ionization cross section followed by the *Lotz* [2.19] semiempirical formula. From [2.17]

This process occurs dominantly at electron energies from threshold up to energies when one-electron ionization of the inner shells is possible;
(*ii*) *indirect* processes: indirect ionization is usually referred to processes that occur via intermediate autoionizing states. Such states can be created by a single or multiple ionization of the inner-shell electrons (*Single Innershell Ionization Multiple Autoionization*, SIIMA)

$$e^- + A^{q+} \to e^- + (A^{(q+1)+})^* + e^- \to e^- + A^{(q+m)+} + me^-, \qquad (2.5)$$

or by excitation of one or more inner-shell electrons into autoionizing states with the energy higher than required to ionize the system which then decays by Auger (autoionization) transitions (*Excitation Multiple Autoionization*, EMA)

$$e^- + A^{q+} \to e^- + (A^{q+})^* \to e^- + A^{(q+m)+} + me^-, \qquad (2.6)$$

for simplicity, the intermediate stages (cascades) in (2.5, 6) are omitted; (*iii*) *resonant* processes occuring mainly with *positive ions* via a few steps: in the first step, one (or more) of the electrons in the target ion is excited by the incident electron which, losing most of the initial kinetic energy, is resonantly captured into highly excited state of the target ion, forming doubly (or multiply) excited autoionizing state. In the second step, this state decays by sequential emission of electrons

$$e^- + A^{q+} \rightarrow (A^{(q-1)+})^{**} \rightarrow (A^{q+})^* + e^- + ... \rightarrow A^{(q+m)+} + (m+1)e^-, \quad (2.7)$$

or by simultaneous emission of additional electrons

$$e^- + A^{q+} \rightarrow (A^{(q-1)+})^{**} \rightarrow A^{(q+m)+} + (m+1)e^-. \quad (2.8)$$

The process (2.7) is called *Resonant Excitation Multiple Autoionization* (REMA) and the process (2.8) is *Resonant Excitation Auto-Multiple Ionization* (REAMI). Certainly, the resonant processes considered can be also referred as indirect processes; the difference is that the resonant processes occur via electron capture similar to dielectronic recombination process.

The roles of indirect and resonant processes in electron impact ionization of neutral atoms and positive ions have been considered in [2.4, 6, 20, 21]. We note that in multiple ionization of highly charged ions, the decay of multiply excited states in the processes (2.5–8) has another competitive channel - radiative decay which has to be taken into account.

Multiple Ionization of Neutral Atoms. At present, there have been reported measurements of MI cross sections σ_m of various atomic targets from Ne up to U at electron energies from threshold up to 2×10^4 eV and the numbers of ejected electrons from $m = 1$ to $m = 13$ (see, e.g., [2.11]). Ionization cross section data on σ_m for rare gases [2.11, 22, 23] have been investigated in more detail as compared to other targets. However, even for these species the experimental data are sometimes not consistent and complete and large discrepancies exist among MI cross sections, in particular for large m.

Relative contribution of the MI cross sections of neutrals is shown in Fig. 2.1 for e + Cu collisions. With the number m of ejected electrons increasing, the cross section σ_m rapidly decreases. Ionization of Mg atoms is an exception: the double-ionization cross section σ_2 is comparable with the single one σ_1 at electron energies E > 1 keV (Fig. 2.2). Besides direct double ionization, σ_2 is strongly enhanced by Auger transition $2p^5 3s^2\,{}^2P \rightarrow 2p^6\,{}^1S + e^-$ in Mg^+ after removal of the inner-shell $2p$-electron at 55.8 eV.

Multiple Ionization of Positive Ions. Experimental data on σ_m for positive ions are not so complete as for neutral atoms and are available mainly for ejection of $1 \le m \le 4$ electrons from ions of rare-gas atoms (He, Ne, Ar, Kr, Xe), alkali (Rb, Cs), metals (Fe, Ni, Mo, Ba, La, Ce, W) and some other

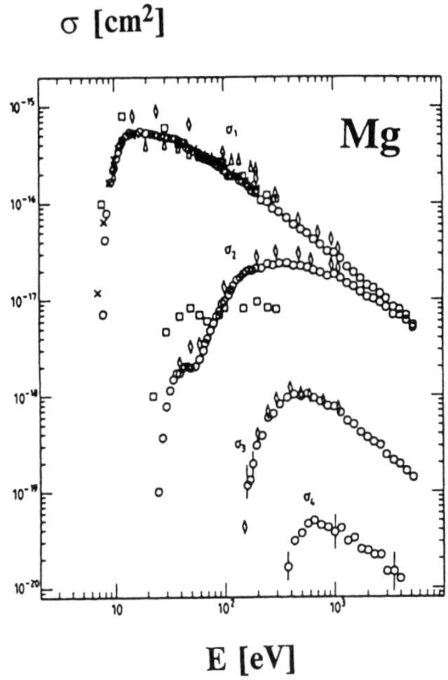

Fig. 2.2. Experimental m-fold ionization cross sections σ_m of Mg by electron impact. Circles – σ_1, σ_2, σ_3 and σ_4, [2.24]; triangles – σ_1, [2.25]; diamonds – σ_1, σ_2 and σ_3, [2.26]; crosses – σ_1, [2.18]; squares – σ_1, σ_2, [2.27]. From [2.24]

elements (Bi, Sb, I) (see [2.6, 11] for references). In general, the properties of MI cross sections for positive ions are similar to those for neutral atoms, except for the fact that in the ionic targets the contribution from direct m-electron ionization [DMI, (2.4)] strongly decreases with the ion charge increasing due to the increased binding energy. The contribution of the inner-shell electron ionization to MI cross sections can be quite significant as demostrated in Fig. 2.3 for double ionization of $Kr^{q+}(3p^6 3d^r)$ ions ($q =$ 9–13, $r =$ 9–5). Experimental data are presented in a linear scale to show the shoulders at around 1.8 keV which are due to the single ionization of the inner $3p$-electrons.

Multiple Ionization of Negative Ions. At present, experimental data on double-ionization cross sections for H^-, C^-, O^- and F^- ions are known only from the literature [2.14]. Double ionization of negative ions results in a positive ion production with two target electrons involved: the binding energy of the first, outer electron is very small (by about one order of magnitude) compared to that of the second electron. The contribution of the inner-shell electron ionization becomes significant for many-electron negative ion systems.

A comparison of experimental MI cross sections for atoms, positive and negative ions with available semiempirical and asymptotic formulas is given in the next sections.

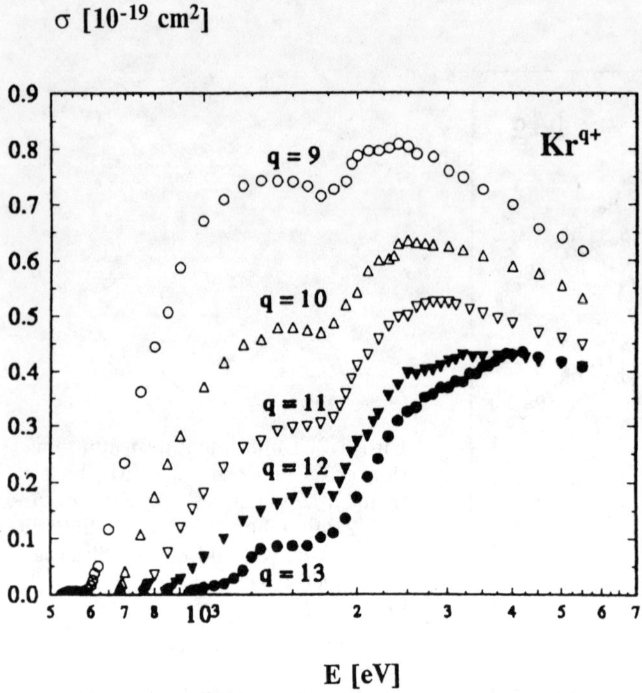

Fig. 2.3. Experimental double-ionization cross sections of $\mathrm{Kr}^{q+}(3p^6 3d^r)$ ions, $q = 9$–13, $r = 9$–5, by *Oualim* et al. [2.28,29]. In the vicinity of the electron energy $E \approx 1.8$ keV there are shoulders on the cross sections caused by contribution from ionization of the inner $3p$-electrons

2.1.2 Semiempirical Formulas. Comparison with Experiment

Semiempirical formulas for m-electron ionization cross sections, $m \geq 2$, of atoms and ions by electron impact have been independently developed under different approaches [2.13,14,30–32]. Usually, the shape of the ionization cross sections is described by different analytical expressions with approximation parameters estimated by fitting the model cross sections to the experimental data. The Bethe-Born type formula [2.13,14] for the m-electron ionization cross section, $m \geq 2$, has the form

$$\begin{aligned} \sigma_m &= 10^{-18}\mathrm{cm}^2 \frac{a(m)N^{b(m)}}{(I_m/\mathrm{Ry})^2}\left(\frac{u}{u+1}\right)^c \frac{\ln(u+1)}{u+1}, \\ u &= E/I_m - 1, \ m \geq 2, \end{aligned} \tag{2.9}$$

where E is the incident electron energy, N is the total number of the target electrons, I_m is the threshold energy given by (2.2), a and b are the fitting parameters and Ry is the Rydberg unit, $1\text{Ry} = 13.606$ eV. The exponent c is determined empirically: $c = 1.0$ for neutrals and $c = 0.75$ for ionic targets.

The fitting parameters $a(m)$ and $b(m)$ evaluated from reliable experimental data for ionization of $2 \le m \le 10$ electrons are listed in Table 2.1. If $m > 10$, the asymptotic values can be used

$$a(m) \approx 1350 m^{-5.7}, \quad b(m) = \text{constant} = 2.00, \quad m > 10, \tag{2.10}$$

obtained by extrapolation of the $a(m)$ and $b(m)$ values for $m \le 10$.

Table 2.1. Fitting parameters $a(m)$ and $b(m)$ in (2.9) for ejection of m electrons from atoms and ions

m	$a(m)$	$b(m)$	m	$a(m)$	$b(m)$
2	14.0	1.08	7	0.021	2.00
3	6.30	1.20	8	0.0096	2.00
4	0.50	1.73	9	0.0049	2.00
5	0.14	1.85	10	0.0027	2.00
6	0.049	1.96			

The parameters $a(m)$ and $b(m)$ are smooth functions of the number of ejected electrons m. The values for the ionization energies I_m can be found from tables of *Lotz* [2.15] and *Carlson* et al. [2.16]; for some neutral atoms the values of I_m, $m \ge 2$, are given in Table 2.2.

Table 2.2. Total number N of the target electrons and ionization energies I_m (in eV) for neutral atoms

Atom	N	I_1	I_2	I_3	I_4	I_5	I_6	I_7	I_8	I_9	I_{10}	I_{11}	I_{12}	I_{13}
He	2	24.6	79											
Ne	10	21.6	63	127	223	349	513	734	986	2109	3405			
Mg	12	7.6	22	103	212	353	539	788	1066	1423	1819	3490	5370	
Ar	18	15.8	43	84	144	219	310	450	602	996	1465	2006	2620	3009
Fe	26	7.9	24	56	114	197	305	438	596	836	1107	1410	1745	2117
Cu	29	7.7	20	77	153	256	386	543	728	940	1179	1446	1818	2226
Ga	31	6.0	26	57	118	214	344	508	707	942	1208	1510	1847	2218
Ge	32	7.9	24	59	105	192	317	479	678	916	1190	1504	1853	2242
Se	34	9.7	33	66	112	183	269	416	607	841	1119	1440	1805	2214
Kr	36	14.0	42	75	128	198	283	400	531	753	1020	1357	1700	2220
Ag	47	7.6	29	70	131	211	310	429	568	728	906	1105	1383	1686
In	49	5.8	24	52	106	184	284	408	555	725	919	1137	1378	1643
Sn	50	7.3	22	53	95	169	267	392	540	713	911	1135	1384	1658
Sb	51	8.6	25	53	98	155	250	372	520	695	896	1123	1378	1660
Te	52	9.0	29	58	98	160	233	353	500	675	878	1109	1365	1653
Xe	54	12.1	35	64	109	169	240	340	450	622	824	1056	1320	1648
Pb	82	7.4	23	55	99	166	254	363	492	642	813	1008	1224	1462
Bi	83	7.3	24	51	97	156	241	350	480	631	804	1000	1219	1462
U	92	6.0	17	36	67	116	185	275	380	500	633	798	979	1202

According to (2.9), the MI cross section σ_m reaches its maximum at $u_m^{\max} \approx 3.2$, i.e.,

$$E_m^{\max} \approx 4.2 I_m, \ \sigma_m^{\max} \approx 0.27 \times 10^{-18}\, a(m) N^{b(m)} (\mathrm{Ry}/I_m)^2\, \mathrm{cm}^2. \quad (2.11)$$

Equation (2.11) gives quite good estimate for the cross section maximum and the corresponding electron energy $E_m^{\max}$. For example, the experimental [2.18] triple-ionization cross section of Bi atoms has the maximum value

$$\sigma_3^{\max} = 2.7 \times 10^{-17} \mathrm{cm}^2 \text{ at } E_3^{\max} = 170\,\mathrm{eV}, \quad (2.12)$$

while (2.11) gives ($N = 83$, $I_3 = 51.3$ eV), respectively:

$$\begin{aligned} \sigma_3^{\max} &= 0.27 \times 6.3 \times (83)^{1.20} (51.3/13.6)^{-2}\, 10^{-18} = 2.40 \times 10^{-17} \mathrm{cm}^2, \\ E_3^{\max} &= 4.2 \times 51.3 = 215\,\mathrm{eV}. \end{aligned} \quad (2.13)$$

A comparison of estimation (2.11) with experimental data on triple and quadruple ionization of neutrals [2.11, 18, 33] is given in Table 2.3.

Table 2.3. Maximum values of triple- and quadruple-ionization cross sections (in cm^2) and the corresponding electron energies (in eV) from (2.11) in comparison with experimental data [2.11, 18, 33] for neutral atoms. The power of 10 for cross section values is given in brackets.

Atom	$\sigma_3^{\max}$ exp.	$E_3^{\max}$ exp.	$\sigma_3^{\max}$ th.	$E_3^{\max}$ th.	$\sigma_4^{\max}$ exp.	$E_4^{\max}$ exp.	$\sigma_4^{\max}$ th.	$E_4^{\max}$ th.
Mg	1.0(-18)	430	5.8(-19)	430	5.0(-20)	650	4.0(-20)	890
Ar	9.1(-19)	570	1.4(-18)	350	1.9(-19)	1150	1.8(-19)	600
Fe	4.3(-18)	130	5.0(-18)	240	8.0(-19)	450	5.4(-19)	480
Cu	3.0(-18)	350	3.0(-18)	320	5.3(-19)	770	3.6(-19)	640
Ga	5.8(-18)	290	6.0(-18)	240	7.4(-19)	560	6.8(-19)	500
Ge	3.5(-18)	185	5.9(-18)	250			7.3(-19)	440
Se	5.1(-18)	195	4.9(-18)	280			8.9(-19)	470
Kr	4.9(-18)	400	3.3(-18)	350	1.5(-18)	500	6.4(-19)	580
Ag	9.4(-18)	180	6.5(-18)	290			1.1(-18)	580
In	9.6(-18)	130	1.2(-17)	220			1.8(-18)	550
Sn	1.7(-17)	165	1.2(-17)	220			2.4(-18)	440
Sb	2.2(-17)	150	1.2(-17)	430			2.3(-17)	400
Te	2.1(-17)	150	1.1(-17)	240			2.4(-18)	410
Xe	2.0(-17)	140	7.5(-18)	300	5.0(-18)	225	1.8(-18)	490
Pb	2.1(-17)	185	2.0(-17)	230			5.2(-18)	410
Bi	2.7(-17)	170	2.4(-17)	220			5.3(-18)	410
U	8.0(-17)	180	5.6(-17)	150	3.3(-17)	200	1.4(-17)	280

In general, the data for ions obtained by different experimental groups are rather consistent and are believed to be relatively accurate. The semiempirical Bethe-Born type formula (2.9) can reproduce quite well the observed

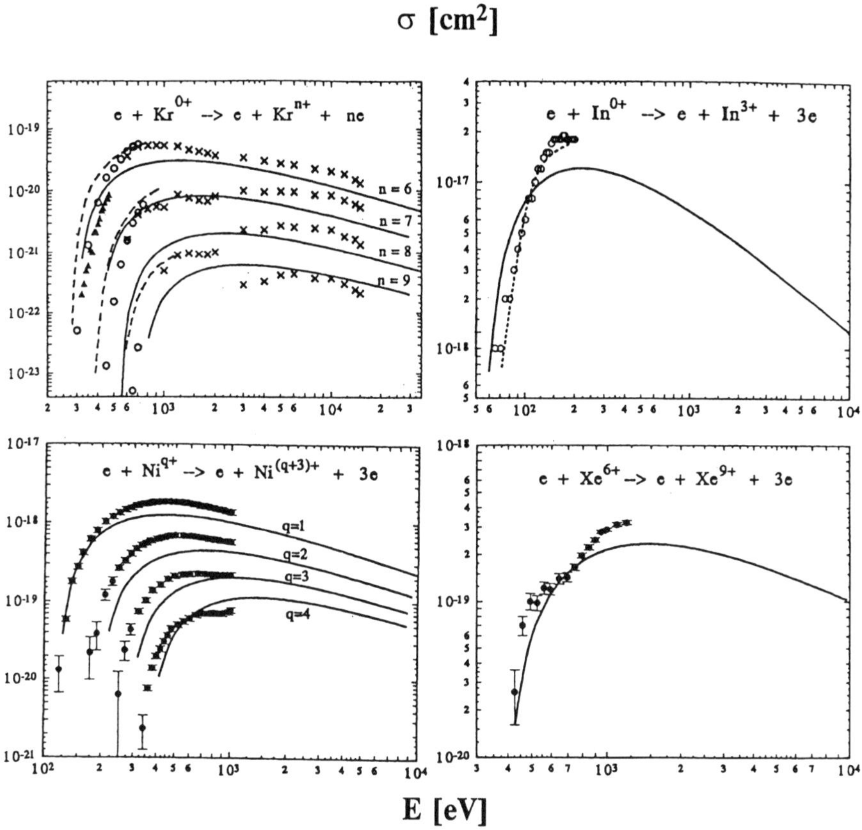

Fig. 2.4. MI cross sections of Kr, In atoms and Ni^{q+}, Xe^{6+} ions. Symbols correspond to experimental data, solid curves are semiempirical cross sections (2.9). Kr: crosses – [2.34], open circles [2.35], triangles – [2.36]; dashed curves – semiempirical formula [2.31] for $m = 6$, 7 and 8; In: open circles – [2.18]; dotted curve – result of semiempirical formalism [2.32]; Ni^{q+}: circles – [2.37]; Xe^{6+}: circles – [2.38]

data for most of the cases except for those where the indirect processes mentioned are important (see next Section). Some experimental MI cross sections are displayed in Fig. 2.4 in comparison with formula (2.9) and other semiempirical formulas, e.g., with results of the scaling formula [2.31] and semiempirical formalism [2.32]. All semiempirical formulas mentioned give approximately the same accuracy within a factor of 2. One of the main advantages of the formula (2.9) is that it gives analytical dependence of σ_m on the ionization potential I_m, the number of ejected electrons m and the total number of the target electrons N. From (2.9) and Table 2.1, the MI cross sections at $m > 6$ are given in the following form

$$\sigma_m \sim N^2 m^{-6} I_m^{-2} F(u), \;\; u = E/I_m - 1, \tag{2.14}$$

where $F(u)$ is the universal function given in (2.9) for all m.

2.1.3 Contribution of Direct and Indirect Processes

Let us consider the contribution from direct and indirect processes in the case of double ionization. In the approximation of the *independent* processes the corresponding cross section can be written in the form

$$\sigma_2 = \sigma_2^{\text{dir}} + \sum_i \sigma_1^{(i)} B_i + \sum_j \sigma_{ex}^{(j)} B_j^{\text{da}}, \;\; E \geq I_2, \tag{2.15}$$

where I_2 is the threshold energy for double ionization, and E is the incident electron energy.

The first term, σ_2^{dir}, is the direct double-ionization cross section related with electron correlation effects (Sect. 2.1.4). The second term is the sum over all hole states i of the target that can contribute to double ionization (ionization-autoionization). Here $\sigma_1^{(i)}$ is the single ionization cross section of the inner-shell electron and B_i is the branching-ratio coefficient given by

$$B_i = \sum_k A_{ik}^a / \Gamma, \tag{2.16}$$

where A_{ik}^a is the autoionization probability for decay channels k and Γ is the sum of the all autoionization A^a and radiative A^r decay probabilities of the state i. The third term in (2.15) describes excitation process of the inner-shell electrons into autoionizing states which can decay by appropriate Auger processes. Here $\sigma_{ex}^{(i)}$ is the single excitation cross section induced by electron impact and B_j^{da} is the double-autoionization branching ratio

$$B_j^{\text{da}} = \frac{\sum_{j'} A_{jj'}^a \sum_f A_{j'f}^a}{(\sum_k A_{jk}^a + \sum_l A_{jl}^r)(\sum_k A_{j'k}^a + \sum_l A_{j'l}^r)}. \tag{2.17}$$

In the case of MI of ionic targets, one also has to include the resonant capture processes to the total cross section in (2.15). As it is seen, calculation even for the double-ionization cross section in the independent process approach faces severe difficulties which strongly increase with increasing the number of ejected electrons [2.6].

As follows from (2.16), only direct double ionization can occur if excitation or ionization channels of the inner-shell electrons are closed energetically. Experimental results obtained by *Müller* et al. [2.39–42] and *Oualim* et al. [2.28, 29] have demonstrated that the contribution of the direct cross sections decrease with increasing the target ion charge and the number of electrons ejected. Therefore, for complex highly charged ions multiple ionization processes are dominated by the indirect processes involving mostly one-electron transitions. It is clearly seen in Fig. 2.3 where experimental double-ionization cross sections σ_2 at $E \approx$ 1.8 keV are caused by the interplay between direct and in direct processes.

Although the reliable theoretical calculations of the direct multiple ionization cross sections σ_m^{dir} are practically absent, the values of σ_m^{dir} can be estimated from experimental data of the total cross sections, σ_m, and reliable one-electron cross sections and branching ratios. For example, the direct triple-ionization cross sections σ_3^{dir} in collisions of electrons with Ar^+ [2.39] and Ar [2.43] targets have been estimated presenting the experimental data in the form (see (2.15))

$$\sigma_3 = \sigma_3^{\mathrm{dir}} + a_{2s}\sigma_{2s} + a_{2p}\sigma_{2p}. \tag{2.18}$$

Here σ_3^{dir} is the direct-ionization cross section (simultaneous ionization of three outer $3p$-electrons); σ_{2s} and σ_{2p} are the single-ionization cross sections of $2s$ and $2p$ electrons, respectively. The coefficients a_{2s} and a_{2p} describe the Auger and Coster-Kronig rates contributing to the total triple-ionization cross section.

The cross section in question is displayed in Fig. 2.5. At electron-impact energies $E < 250$ eV, only the direct ionization is possible because the channels for single ionization of the inner $2s$ and $2p$ electrons are closed (the corresponding binding energies $I_{2p} = 250$ eV and $I_{2s} = 320$ eV). Therefore, the first maximum ($E \simeq 200$ eV) corresponds to direct ionization of three outer electrons, while the second one ($E \simeq 700$ eV) is the result of Auger decay of Ar^+ ion from autonizing state with a vacancy in the L-shell. Especially, for Ar, L-shell ionization is also found to be important for the production of doubly and triply charged ions [2.44] in collisions with protons and antiprotons.

In practice, however, MI processes may occur in much more complicated way than described by (2.15) which can be seen in an example of triple ionization of Xe^{6+} ions (Fig. 2.4). The measured threshold energy E_{th} coincides with the minimal ionization energy $I_3 = 374$ eV to ionize three outermost electrons

$$\mathrm{e}^- + \mathrm{Xe}^{6+}(4p^6 4d^{10} 5s^2) \to \mathrm{e}^- + \mathrm{Xe}^{9+}(4p^6 4d^9) + 3\mathrm{e}^-, \tag{2.19}$$

which is a three-electron process. The direct-ionization cross section for the single inner-shell electron is expected to be quite small because the charge of the target is large. The estimated minimum energy to induce a single-electron process is $\Delta E_{3d} \approx 614$ eV which is required to excite the inner $3d$-electron leading to production of Xe^{9+} ions. However, experiment shows a large onset of the cross section between 400 and 600 eV that can

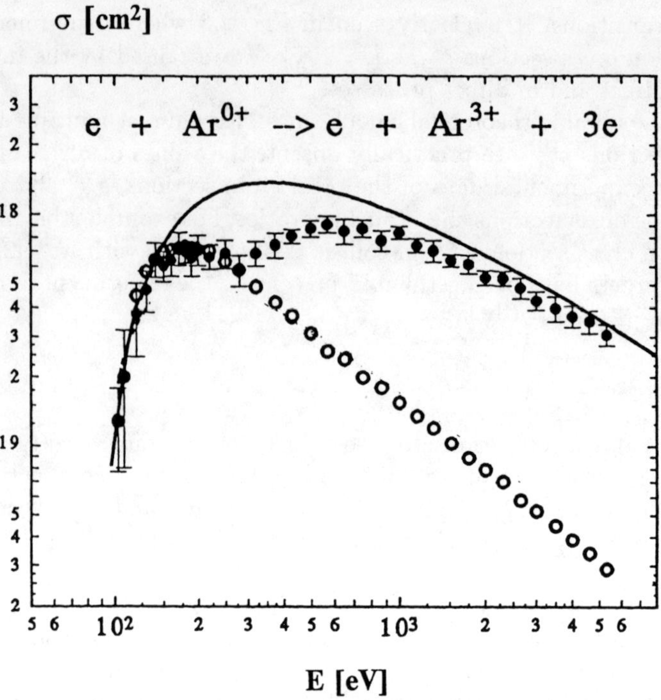

Fig. 2.5. Triple-ionization cross section of Ar atoms. Solid circles – experiment [2.24], solid curve – semiempirical formula (2.9), open circles – estimated [2.43] direct ionization cross section $\sigma_3^{\rm dir}$ of three outer $3p$-electrons with a_{2s}= 0.84 and a_{2p}=0.26 in (2.18)

be explained only by other MI process, e.g., direct (double) ionization of $4p$ and $4d$ electrons followed by Auger decay

$$\mathrm{e}^- + \mathrm{Xe}^{6+}(4p^6 4d^{10} 5s^2) \to \mathrm{e}^- + \mathrm{Xe}^{8+}(4p^5 4d^9 5s^2) + 2\mathrm{e}^- \to$$
$$\mathrm{e}^- + \mathrm{Xe}^{9+}(4p^6 4d^9) + 3\mathrm{e}^-, \qquad (2.20)$$

with the estimated threshold energy $I_2(4p{+}4d) \approx 410$ eV. A further increase of the cross section at $E > 750$ eV is caused by (inner-shell) ionization-autoionization of the $3d$-electron leading to net triple ionization after two Auger decays:

$$\mathrm{e}^- + \mathrm{Xe}^{6+}(3d^{10}4s^2 4p^6 4d^{10} 5s^2) \to \ \mathrm{Xe}^{7+}(3d^9 4s^2 4p^6 4d^{10} 5s^2) + 2\mathrm{e}^- \to$$
$$\mathrm{Xe}^{8+}(3d^{10} 4s 4p^6 4d^{10} 5s) + 3\mathrm{e}^- \to \mathrm{Xe}^{9+}(3d^{10} 4s^2 4p^6 4d^9) + 4\mathrm{e}^-. \quad (2.21)$$

2.1.4 Double Ionization of Two-Electron Targets. Correlation Effects

From a theoretical viewpoint, even a double ionization process

$$e^- + A \to e^- + A^{2+} + 2e^-, \tag{2.22}$$

is a quite complicated reaction because of the presence of four charged particles in the final channel interacting through the long-range Coulomb potential. There are no sophisticated calculations of the integrated double-ionization cross sections so far. (The behavior of the differential double-ionization cross sections has been considered in several papers [2.45–48].) A few attemps have been made to predict double-ionization cross section behavior for electron–atom and electron–ion collisions [2.49–52]. However, for one or another reason, these calculations are unable to provide a satisfactory description of the experimental data and therefore can not be used for applications.

Some of the semiempirical formulas have been already discussed in Sect. 2.1.2. In most cases, these formulas can reproduce quite well the observed data (within a factor of 2) except for those where the indirect processes are important. Some of the double-ionization cross sections are reproduced in Fig. 2.6.

Double ionization of He-like atoms seems to be the simplest multielectron process. It strongly depends on the electron correlation effects which are of great importance for systems with the low nuclear charge, such as H^-, He, Li^+ and Be^{2+}. Because of the strong correlation between target electrons, even in the two-electron targets, double ionization occurs in a quite complicated way via so-called *shake-off* and *two-step* mechanisms [2.44, 57]. Introduction of these mechanisms is quite helpful for understanding double-ionization processes at high velocities of the incident particles.

A sudden single ionization of the target by high-energy electron usually leads to ejection of one (fast) electron. Due to the electron correlations, this causes a sudden change of the (nuclear) effective charge and therefore may lead to additional ionization. This process is called shake-off in the sudden approximation. In high-energy impact ionization, the ratio of double-to-single cross sections becomes constant and independent of the projectile. The shake-off process is usually described within the first Born approximation [2.57].

An ejected electron, if it has sufficient kinetic energy, can collide with the second electron resulting in double ionization. This two-step process (TS-1) involves only one projectile-target-electron interaction. TS-1 process is important at low and high energies. At low velocities, the incident particle can also collide with two target electrons successively and also result in double ionization (TS-2). TS-2 decreases with the velocity increasing. In

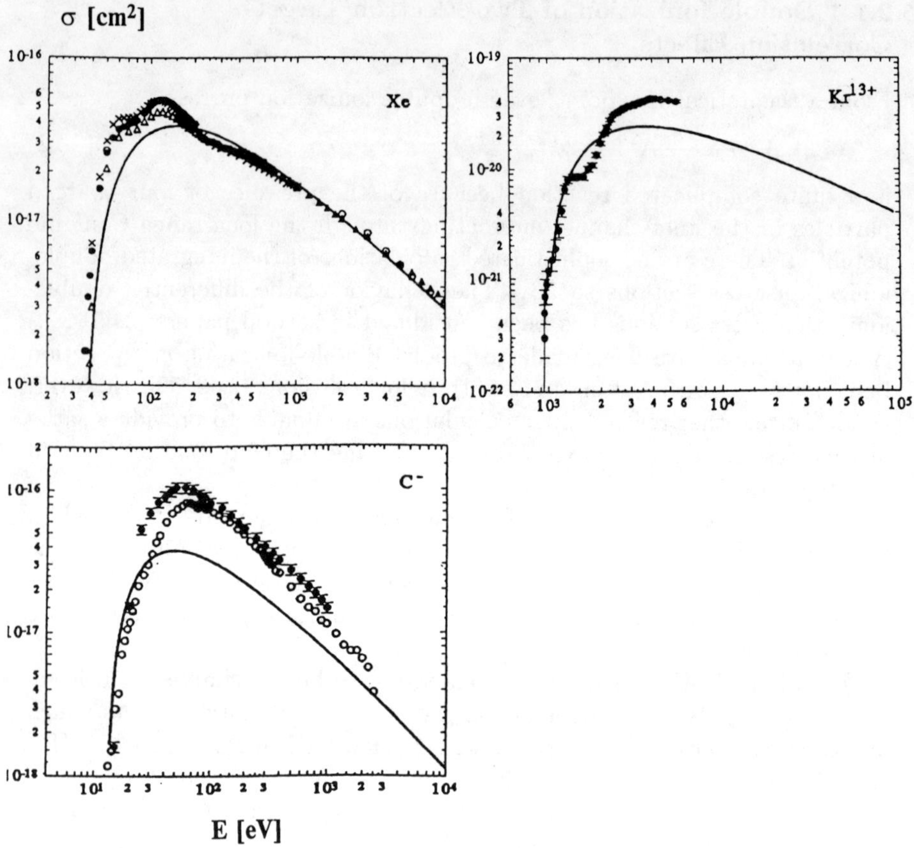

Fig. 2.6. Double-ionization cross sections of Xe, Kr^{13+} and C^-. Symbols are experimental data and solid curve is (2.9). Xe: solid triangles – [2.53], open circles – [2.54], solid circles – [2.55], crosses – [2.22], open triangles – [2.35]; Kr^{13+}: circles – [2.28]; C^-: solid circles – [2.56], open circles – [2.14]

principle, the double ionization process can occur due to the interference between shake-off and TS-2 processes. The interference in double ionization is mainly due to the contribution between two close collisions, TS-1 and TS-2. At any velocity, the dynamic correlation is important in charged-particle impact since the first ionized electron moves with relatively low velocity and, consequently, has ample time to interact with the other target electron of He-like target.

Using this clear interpretation for double ionization of He atom, it is possible to find an asymptotic behavior and a scaling law for the cross sections σ_2 for He-like ion systems. Following the results obtained by *Andersen* et al. [2.44] for double ionization of He by protons, the asymptotic cross section at high relative velocities of the projectile can be presented in the form:

$$\begin{aligned}\sigma_2(v) &= \sigma_{dis}(v) + \sigma_{cl}(v) \\ &= \frac{A}{(v/v_0)^2}\ln(Bv/v_0) + \frac{C}{(v/v_0)^2} + \frac{D}{(v/v_0)^4}, \quad v \gg v_0,\end{aligned} \tag{2.23}$$

representing the contributions from the distant (*dis*, logarithmic term) and close (*cl*, two other terms) collisions, respectively. The last term in (2.23) corresponds to contribution from the two-step mechanism (TS-2). Here v is the relative (incident electron) velocity and v_0 is the atomic unit of velocity ($2.2 \times 10^8 \mathrm{cm\,s^{-1}}$). The first (Bethe logarithmic) term in (2.23) has a purely dipole character, while the second and third terms define other multipole interactions of the incident electron with the target. The dipole part σ_{dis} is related to the double-photoionization cross section $\sigma_2(\omega)$ of the target according to the Weizsäcker-Williams method of virtual quanta (see [2.58] and Sect. 4.1):

$$\sigma_{\mathrm{dis}}(v) = \int_{I_2}^{\infty} Q(\omega, v)\sigma_2(\omega) d\hbar\omega, \tag{2.24}$$

where I_2 is the double-ionization potential and $Q(\omega, v)$ is the intensity of the frequency spectrum caused by the dipole interaction of the incident particle [2.59].

For the high-energy double-ionization cross section of He by electron and proton impact, *Andersen* et al. [2.44] obtained the following constants on the basis of the quantum mechanical consideration and recommended data on $\sigma_2(\omega)$ for He:

$$\begin{aligned}A &= 6.7 \times 10^{-19}\,\mathrm{cm}^2, \quad B = 0.53, \\ C &= 2.36 \times 10^{-18}\,\mathrm{cm}^2, \ D = 1.85 \times 10^{-16}\,\mathrm{cm}^2.\end{aligned} \tag{2.25}$$

Using the scaling relation between double-photoionization cross sections for He-like systems [2.60] and the results of the papers [2.14, 61], one has the following *scaling law* for double-ionization cross sections $\sigma_2(v)$ of two-electron systems by electron impact:

$$Z_{\mathrm{eff}}^6 \sigma_2(v) = f(v^2/Z_{\mathrm{eff}}^2), \ Z_{\mathrm{eff}} = Z - 5/16, \tag{2.26}$$

where Z is the nuclear charge of the target and $f(x)$ is a certain universal function. For highly charged positive ions one has $Z_{\mathrm{eff}} \approx Z$. The scaling $\sigma_2(v) \approx Z^{-6}$ in (2.26) is similar to that obtained by *Ford* and *Reading* [2.62] for proton impact. The asymptotic constants A, B, C, D for an arbitrary two-electron system can be found from (2.23, 26)

$$\begin{aligned}A &= 6.70\beta^4, \ B = 0.53\beta, \ C = 2.36\beta^4, \ D = 3.34\beta^2, \\ \beta &= Z_{\mathrm{eff}}^{\mathrm{He}}/Z_{\mathrm{eff}} = 27/(16Z_{\mathrm{eff}}).\end{aligned} \tag{2.27}$$

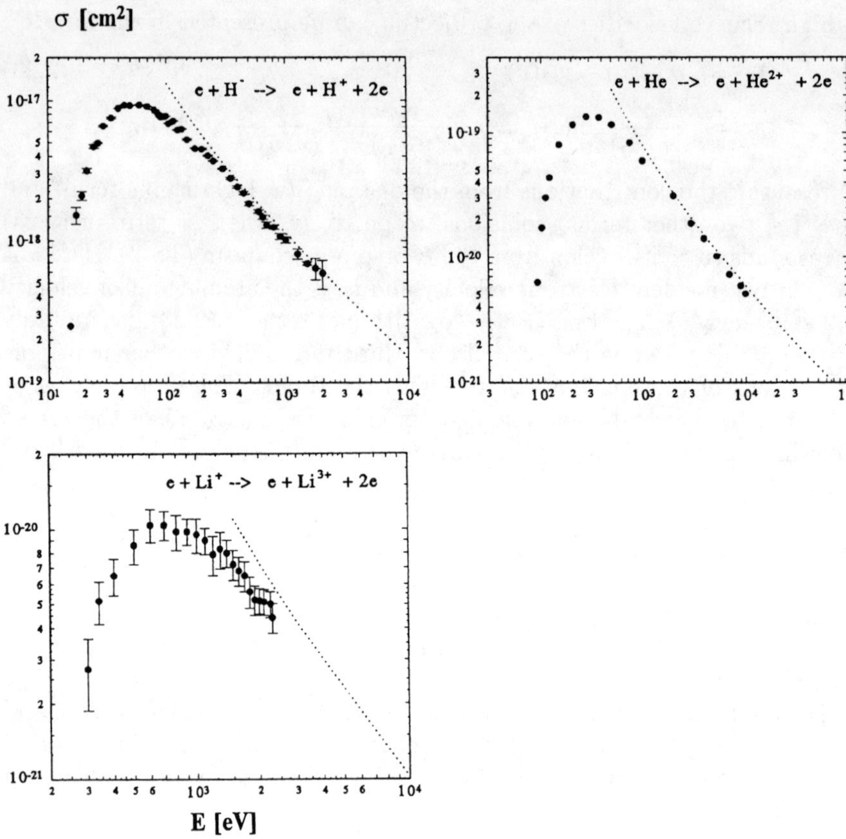

Fig. 2.7. Double-ionization cross sections of He, H^- and Li^+ systems. Solid circles – experimental data: [2.63–65], respectively. Dotted curves – (2.23) with constants A, B, C, D from Table 2.4

The constants A, B, C and D for He, H^-, Li^+ and Be^{2+} are listed in Table 2.4. The asymptotic double-ionization cross sections of He, H^- and Li^+ by electron impact calculated with the help of (2.23) and Table 2.4 are shown in Fig. 2.7. To our knowledge, there is no published experimental data for He-like ions heavier than Li^+.

We note that the ratio of the observed double-to-single ionization cross sections in H^- ions [2.64] at asymptotically high electron energies, (2.3–4.0)$\times 10^{-3}$, seems to be very similar to that of He atoms, 2.6×10^{-3} [2.66], although there are still significant uncertainties. Therefore, to test the theories it would be interesting to remeasure accurate ratios of the cross sections but not necessarily their absolute values.

Table 2.4. Parameters A, B, C and D for asymptotic double-ionization cross sections of two-electron systems, (2.23). From [2.14]

Atom/ion	A, 10^{-19} cm^2	B	C, 10^{-18}cm^2	D, 10^{-16}cm^2
He	6.70	0.53	2.36	1.85
H^-	236	1.15	25.1	3.34
Li^+	1.04	0.33	0.366	0.729
Be^{2+}	0.294	0.24	0.103	0.387

Table 2.5. Experimental ratios $R_m = \sigma_m/\sigma_1$ for m-electron ionization in rare-gas atoms at different relativistic energies. From [2.66]

Atom	Ratio	5 keV [2.34]	14 keV [2.53]	40 MeV [2.66]
He	R_2	3.40×10^{-3}	2.78×10^{-3}	2.6×10^{-3}
Ne	R_2	2.70×10^{-2}	2.47×10^{-2}	2.95×10^{-2}
	R_3	1.57×10^{-3}	1.43×10^{-3}	2.2×10^{-3}
Ar	R_2	5.45×10^{-2}	5.45×10^{-2}	6.28×10^{-2}
	R_3	1.38×10^{-2}	1.38×10^{-2}	1.83×10^{-2}
	R_4	2.35×10^{-3}	2.37×10^{-3}	2.82×10^{-3}
	R_5	2.93×10^{-4}	2.69×10^{-4}	3.6×10^{-4}
Kr	R_2	8.08×10^{-2}	9.45×10^{-2}	1.17×10^{-1}
	R_3	4.65×10^{-2}	5.88×10^{-2}	7.44×10^{-2}
	R_4	1.11×10^{-2}	1.40×10^{-2}	1.75×10^{-2}
	R_5	3.69×10^{-3}	4.26×10^{-3}	-
	R_6	9.98×10^{-4}	1.33×10^{-3}	1.7×10^{-3}
	R_7	3.47×10^{-4}	5.32×10^{-4}	6.0×10^{-4}
Xe	R_2	1.53×10^{-1}	1.71×10^{-1}	2.64×10^{-1}
	R_3	6.39×10^{-2}	6.86×10^{-2}	1.12×10^{-1}
	R_4	2.14×10^{-2}	2.32×10^{-2}	3.49×10^{-2}
	R_5	6.41×10^{-3}	7.84×10^{-3}	1.49×10^{-2}
	R_6	3.02×10^{-3}	4.15×10^{-3}	7.18×10^{-3}
	R_7	1.49×10^{-3}	2.15×10^{-3}	2.6×10^{-3}
	R_8	7.59×10^{-4}	1.11×10^{-3}	1.4×10^{-3}

2.1.5 Ionization by Relativistic Electrons

Beside the purely fundamental interest in MI of atoms, this process is very effective in production of slow multiply charged ions in single collisions of fast heavy ions (Sect. 2.2) or relativistic electrons. In [2.66], the relative abundances F_m of the resulting A^{m+} ions have been measured in collisions of relativistic 20–50 MeV electrons with rare-gas atoms, A = He, Ne, Ar, Kr and Xe, using a time-of-flight technique. The charge states up to $m = 8$ for Xe have been observed with F_m values being constant in the investigated energy range but strongly decreasing with m increasing. In Fig 2.8 a comparison of the measured relative fraction F_m as a function of m for

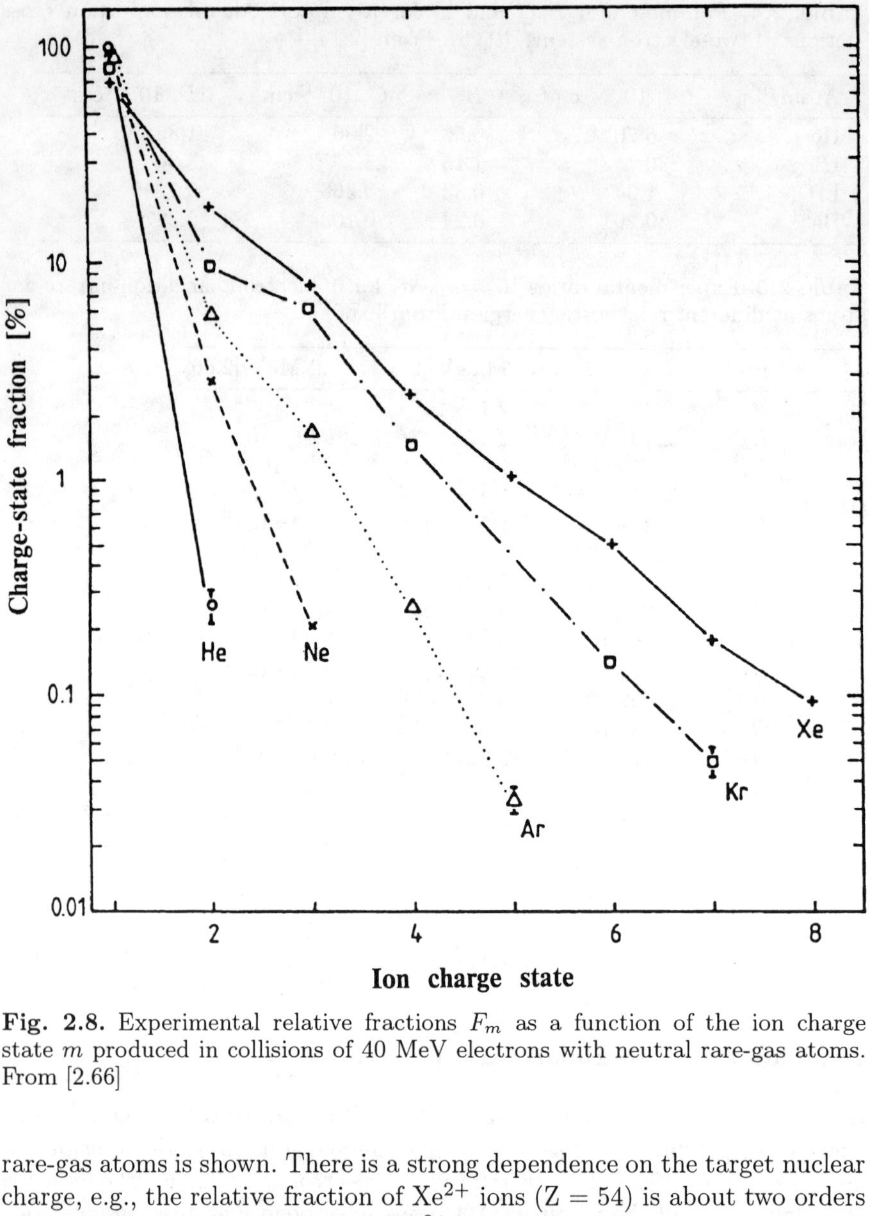

Fig. 2.8. Experimental relative fractions F_m as a function of the ion charge state m produced in collisions of 40 MeV electrons with neutral rare-gas atoms. From [2.66]

rare-gas atoms is shown. There is a strong dependence on the target nuclear charge, e.g., the relative fraction of Xe^{2+} ions (Z = 54) is about two orders of magnitude higher than that of He^{2+} ions (Z = 2).

Since the measured relative abundances F_m are constant in the 20–50 MeV energy range, the ratios of the corresponding MI cross sections to the single one, $R_m = \sigma_m/\sigma_1$, should remain also independent of the incident electron energy that already was found at lower energies in the previous

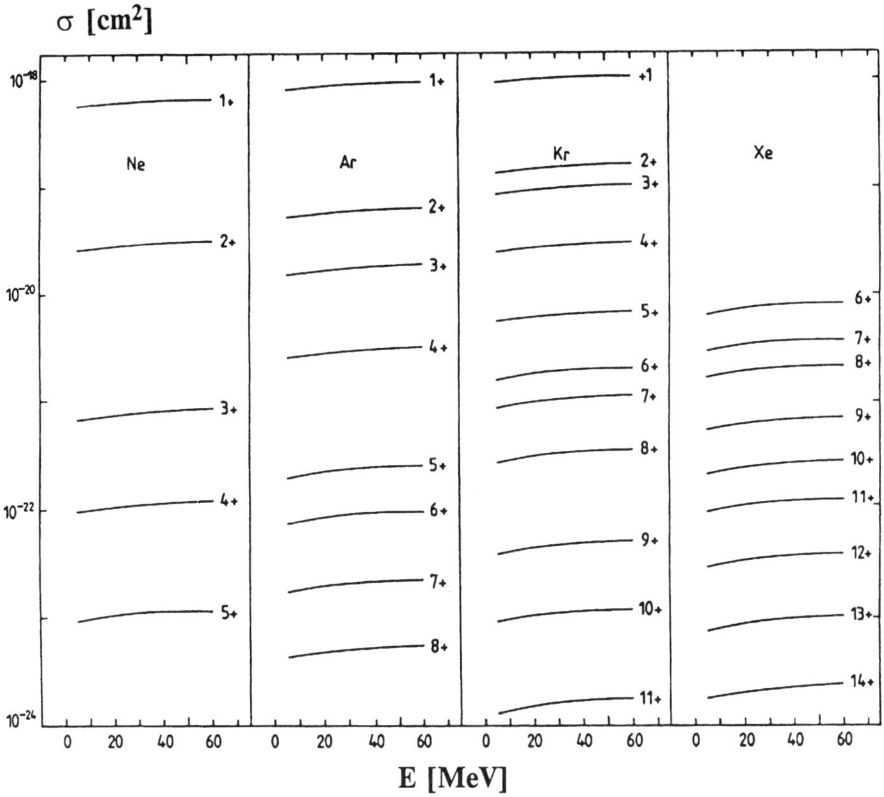

Fig. 2.9. Calculated MI cross sections in collisions of relativistic electrons with inert gases for production of A^{m+} ions. From [2.66]

experiments [2.34, 53]. In Table 2.5 the experimental ratios R_m are given for rare-gas atoms. It is seen that the contribution of MI cross sections to the total one, i.e., the value

$$\sum_{m\geq 2} \sigma_m / \sum_{m=1} \sigma_m = \sum_{m\geq 2} R_m / \left(1 + \sum_{m\geq 2} R_m\right), \tag{2.28}$$

strongly increases with the nuclear charge of the target and constitutes approximately 0.26% for He, 3% (Ne), 8% (Ar), 20% (Kr) and 30% (Xe).

The calculated absolute MI cross sections σ_m for Ne, Ar, Kr and Xe atoms at 5–60 MeV electron energies [2.66] are shown in Fig. 2.9. The σ_m values have been estimated by

$$\sigma_m = \sum_t a_t(m)\sigma_{1t}, \tag{2.29}$$

using the compiled experimental and theoretical data on single-ionization cross sections σ_{1t} from the inner t-shell (K, L or M) of various atomic targets [2.67] and a semiempirical scaling for σ_{1t} at relativistic energies

$$\sigma_{1t} = 10^{-24}\mathrm{cm}^2\, N_t(I_t/\mathrm{keV})[144.6\,\ln(E/I_t) + 83]. \tag{2.30}$$

Here N_t and I_t are the number of equivalent electrons and the binding energy of the inner shell t, respectively. The shake-off probabilities $a_t(m)$, describing the relaxation by Auger transitions after the sudden production of the inner-shell vacancies, have been taken from the work by *Carlson* et al. [2.68] who evaluated them from photoionization data. Figure 2.9 reflects a weak (logarithmic) energy dependence of σ_m on the electron energy E as given by (2.30). Theoretical fractions F_m obtained from cross sections σ_m estimated in this way are in rather good agreement with the measured F_m values.

At ultrarelativistic energies, the single- and double-ionization cross sections of He are estimated to be [2.44]:

$$\sigma_1 = 1.98q^2\ln(248\gamma)10^{-20}\mathrm{cm}^2, \tag{2.31}$$

$$\sigma_2 = 3.6q^2\ln(1470\gamma)10^{-23}\mathrm{cm}^2, \tag{2.32}$$

$$\gamma = \left[1-(v/c_\mathrm{o})^2\right]^{-1/2},\ \gamma \gg 1, \tag{2.33}$$

where q is the projectile charge. For $E \approx 40$ MeV, σ_2 from (2.32) is about 20% lower than that given in Fig. 2.9.

2.2 Multielectron Ionization by Positive Ions

2.2.1 General

Multiple target ionization in a single collision of a fast highly charged positive ion with a many-electron atom occurs with large cross sections [2.69,70]. Therefore, it significantly contributes to the energy loss and straggling of ions in gases [2.71], plasmas [2.72] and solid matter [2.73, 74]. Also, fast heavy-ion beams are very efficient for producing highly charged, very slow (a few eV) recoil ions which can be used as a spectroscopic source of VUV and x-ray radiation.

The processes arising in ion-atom collisions

$$\mathrm{A}^{q+} + \mathrm{B} \rightarrow \mathrm{A}^{q'+} + \mathrm{B}^{m+} + (q'-q+m)\mathrm{e}^-, \tag{2.34}$$

are much more complicated as compared to electron–atom collisions because of the complex atomic structure of the incoming projectile. In the case of ion–atom collisions there are three main multielectron processes distinguished in experiment:

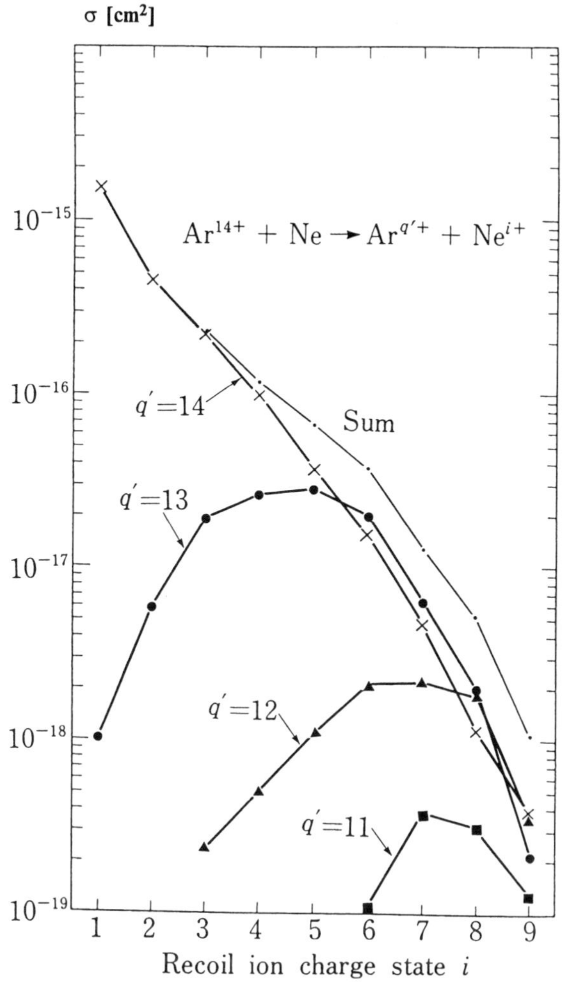

Fig. 2.10. Experimental partial cross sections for recoil-ion production in collisions of 1.05 MeV/u Ar^{14+} ions with Ne. Sum denotes the total cross section summed over pure ionization ($q' = 14$) and transfer ionization ($q' = 11-13$) cross sections, respectively. From [2.75]

i) *pure multiple ionization*, $q' = q$,
ii) *ionization* with *capture* (or transfer ionization), $q' < q$,
iii) *ionization* with *loss* (or loss ionization), $q' > q$.

As a rule, the processes i) and ii) give the main contribution to the total ionization cross section. Typical behavior of the experimental partial cross sections $\sigma^m_{q,q'}$ for reactions (2.34) are shown in Fig. 2.10 for 1.05 MeV/u Ar^{14+} ions colliding Ne atoms.

Double ionization of He by protons and highly charged particles is the simplest but the most fundamental multiple ionization process because of a strong influence of the correlation effects. In addition to the electron–electron correlation, the heavy, highly charged projectile may interact also with two target electrons promoting them into continuum [2.76]. Quantum-mechanical approaches like the Forced Impulse Approximation (FIM) [2.77, 78] or the nonstationary Volkov-Keldysh approximation [2.79–81] correctly describe and predict double-electron ionization cross sections as well as the ratios of double to single ionization cross sections [2.82]. Figure 2.11 shows calculated scaled double-ionization cross sections of He by positive ions in comparison with experimental data.

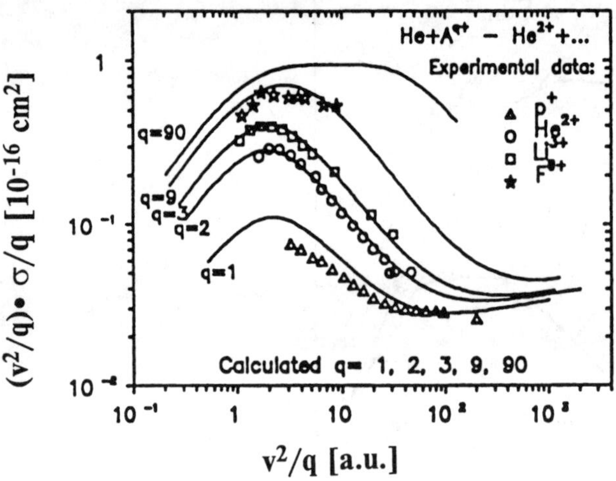

Fig. 2.11. Scaled cross sections of double-electron detachment from He in collisions with protons and highly charged ions as a function of scaled velocity v^2/q: symbols – experiment [2.83–85], solid curves – Volkov-Keldysh calculations. From [2.81]

Useful information on the charge-state distributions and partial-ionization cross sections of the recoil ions correlated with the projectile final charge states is provided applying the coincidence technique for projectile and recoil ions [2.75, 86–96] (Chap. 1). Reliable experimental data on multielectron cross sections are quite scarce although they are known for the projectiles from H^+ up to U^{92+} in the 1–120 MeV/u energy range for ionization of m = 1–35 electrons mainly in collisions with rare-gas atoms [2.97–112].

Analysis of the charge-state distributions of slow recoil ions in ion–atom collisions has shown that the main channels for mutiple target ionization are:

i) direct multiple ionization of outer-shell electrons,
ii) multiple electron capture,
iii) single and multiple electron capture with simultaneous ionization,
iv) inner-shell ionization followed by Auger electron emission,
v) shake-off mechanisms when a sudden ionization of a single electron can lead to the ensuing readjustment to the new core potential which may cause one or more of the remaining electrons to be ejected.

These processes are similar to multiple electron processes arising in collisions of atoms with electrons (Sect. 2.1) and photons (Chap. 4).

The charge-changing cross section $\sigma_{q,q'}$ of the projectile from the charge q to q' including all the ionization states m of the target is given by

$$\sigma_{q,q'} = \sum_{m\geq 1} \sigma^m_{q,q'} . \tag{2.35}$$

The total ionization cross section for target atoms is defined as a sum of the pure m-fold ionization cross sections

$$\sigma_{tot} = \sum_m \sigma_m . \tag{2.36}$$

Usually, the cross sections σ_m for producing a recoil ion in the charge state m is obtained by normalizing the measured charge-state fraction to the *net* ionization cross section σ_+

$$\sigma_+ = \sum_m m\,\sigma_m , \tag{2.37}$$

which is proportional to the total charge of the projectile ions measured in experiment.

The average charge $< m >$ of the target ions is obtained from

$$< m >_{q,q'} = \sum_{m\geq 1} m\sigma^m_{q,q'}/\sigma_{q,q'} , \tag{2.38}$$

and the average charge of the projectile from,

$$< z > = \sum_z zF_z , \quad z = q' - q , \tag{2.39}$$

respectively, where F_z is the fraction of the projectile which can be determined from the experimental charge distributions.

2.2.2 Theory and Experiment

Theoretical and experimental aspects of pure multiple ionization processes

$$\mathrm{A}^{q+} + \mathrm{B} \rightarrow \mathrm{A}^{q+} + \mathrm{B}^{m+} + m\mathrm{e}^- \tag{2.40}$$

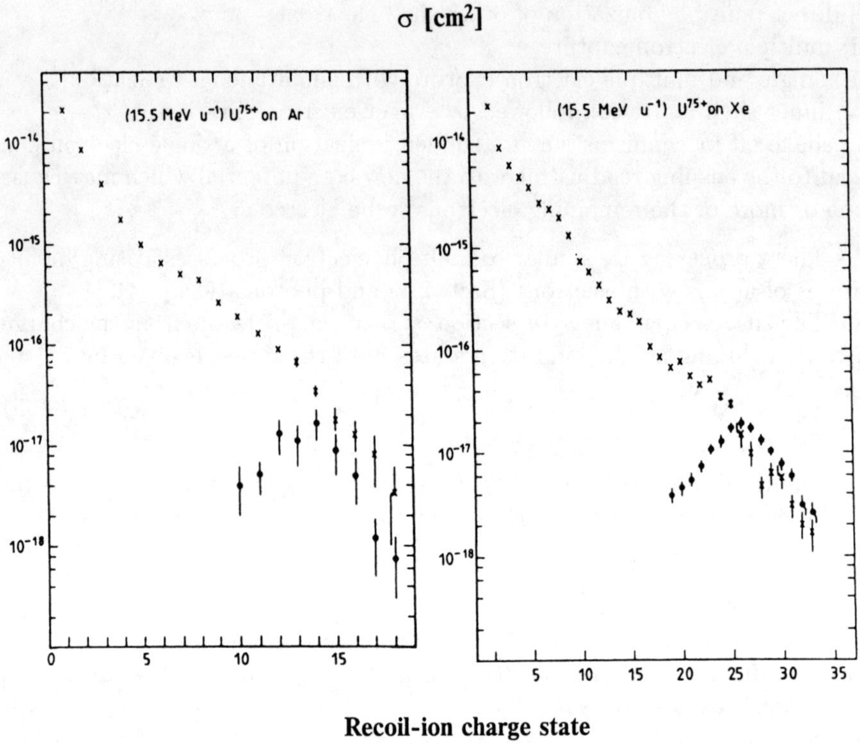

Fig. 2.12. Experimental pure ionization cross sections [2.70] for recoil ion production in collisions of 15.5 MeV/u U^{75+} ions with Ar and Xe: crosses – direct ionization, circles – electron capture

were considered in [2.5, 10, 86, 113, 114]. Typical examples of the pure ionization cross sections are displayed in Figs. 2.12 and 2.13. One can see that the cross sections are very large and show little effects from the target atomic structure. Experiments and theories showed that even highly charged recoil ions are mainly produced at impact parameters considerably larger than the mean shell radius of the ejected electrons.

In principle, the problem of multiple ionization should be solved within a framework of the quantum-mechanical many-body treatment, for example, Time-Dependent Hartree-Fock approximation (TDHF). For more than two particles, this problem faces significant difficulties. However, if the relative velocity of colliding particles is much larger than the electron orbital velocities, then one can use the semiclassical treatment of many particle dynamics [2.115–117].

Most of the theories developed for multielectron processes in collisions of energetic ions with atoms and molecules mainly rely on the classical or

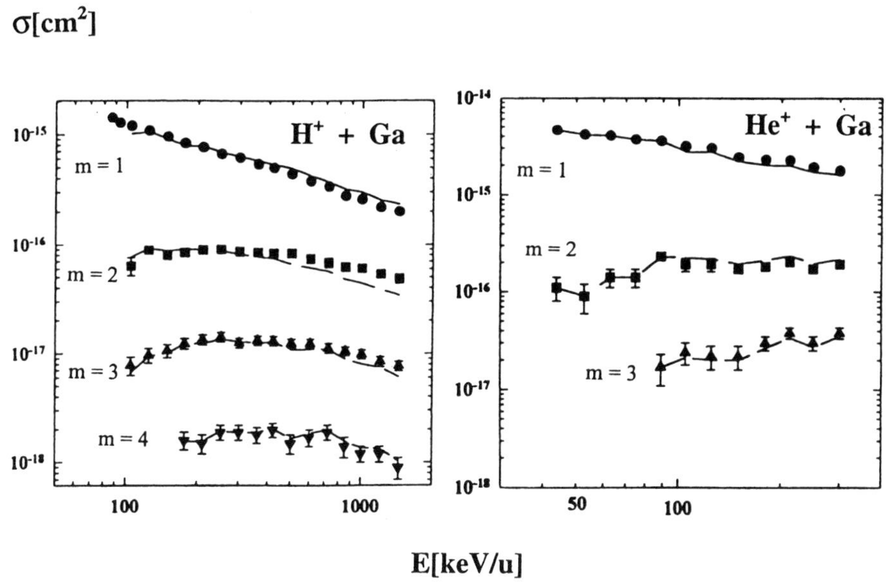

Fig. 2.13. Experimental pure ionization cross sections [2.109] for recoil ion production in collisions of H^+ and He^{2+} ions with Ga: symbols – experimental data, dashed curves – crosses – semiempirical description, see below

semiclassical approximations other than the quantum method. Semiclassical approximation is based on the impact parameter ρ representation which is valid if

$$2Z_pZ_Te^2/\hbar v \gg 1, \tag{2.41}$$

where Z_p and Z_T are the projectile and target charges, respectively, and v is their relative velocity. This condition means that the collision time is shorter than the characteristic orbiting time and that it is satisfied approximately at projectile energies $E_p > 1$ MeV/u. A projectile is usually considered as a classical particle with a straight-line trajectory $\mathbf{R}(\mathrm{t}) = \boldsymbol{\rho} + \mathbf{v}\mathrm{t}$ where ρ is the impact parameter.

Independent Particle Model. The basic approximation used for description of multielectron processes in ion–atom collisions is based on so-called Independent Particle Model (IPM). In this approach, the target electrons are treated independently from each other and electron–electron correlation effects are neglected [2.118–120]. IPM is the most easily applied procedure also used for electron-atom collisions [2.120, 121].

Within the framework of IPM, the probablity P_m for removal of m electrons from the target shell with N electrons is described by the *statistical binomial* distribution

$$P_m = C_N^m P_s^m (1 - P_s)^{N-m}, \quad C_N^m = \frac{N!}{m!(N-m)!}, \tag{2.42}$$

where C_N^m is the binomial coefficient and P_s is the single-electron removal probablity usually calculated within the first-order approximation.

The corresponding cross section for removal of m electrons is then given by

$$\sigma_m(\rho, v) = 2\pi \int_0^\infty C_N^m [P_s(\rho, v)]^m [(1 - P_s(\rho, v)]^{N-m} \rho d\rho. \tag{2.43}$$

If the target consists of a few active shells, then using the single-electron ionization probability for each active shell, one forms products of these and sums all combinations that lead to the ionization with a given charge state. Distributions of the probabilities over the target shells can be found in [2.120,122,123].

For ions with low stages of ionization, the IPM is quite valuable and allows the use of the different theoretical approaches for the single probability P_s such as the Plane Wave Born Approximation (PWBA), semiclassical approximation and Classical-Trajectory Monte Carlo (CTMC) calculations. Though the computation is relatively simple, the IPM suffers from some serious defects, e.g., it ignores that the ionization potential changes considerably with the charge state which leads to wrong probabilities for high stages of ionization. Also, the use of IPM for collisions with molecules [2.124] require additional approximations which practically ignore the molecular structures. However, the IPM model automatically preserves the unitarity of the transition probabilities.

Some theories describing MI processes calculate the ρ-dependence of single-ionization probabilities $P_s(\rho)$ and then make use of the IPM to obtain MI probability and total cross sections.

Born approximation. IPM based on the Born aproximation has been developed in [2.125–127]. In this model the single-particle probability amplitude has the form:

$$P_s = | a_B |^2, \; a_B = \left\langle \mathrm{e}^{i\mathbf{k_f R}} \Phi_f(\mathbf{r}) \left| \frac{Z_p}{R} - \frac{Z_p}{| \mathbf{r} - \mathbf{R} |} \right| \mathrm{e}^{i\mathbf{k_i R}} \Phi_i(\mathbf{r}) \right\rangle, \tag{2.44}$$

where R is the internuclear separation, Z_p is the projectile charge, $k_{i,f}$ is the momentum of the system and $\Phi_{i,f}(\mathbf{r})$ is the corresponding the wave functions before and after collision. Usually, some additional approximations are made, e.g., the screening effects are treated statically only, wave functions do not depend on **R** and so on. Figure 2.14 shows the calculated multi-electron probabilities as a function of the impact parameter for 35 MeV/u Kr^{30+} ions colliding with Ne atoms; both approximations, Born IPM and statistical model (2.42), are in a quite good agreement.

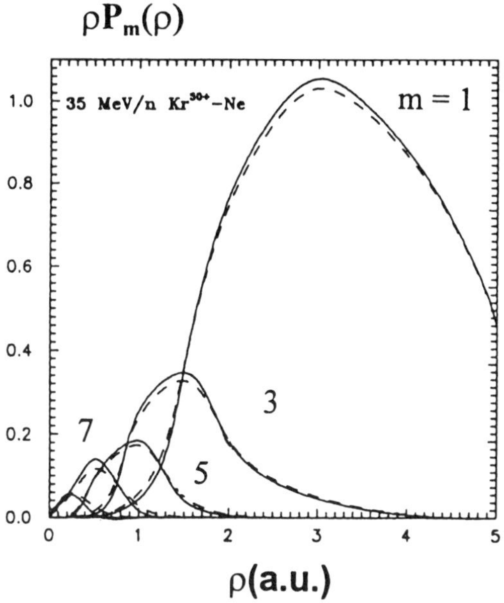

Fig. 2.14. Calculated probabilities for m-electron ionization in collisions of 35 MeV/u Kr^{30+} ions with Ne: solid curves – Born-IPM model, dashed curves – statistical model. From [2.125]

CTMC approach. The CTMC was developed by *Olson* et al. [2.103, 128–130]. The CTMC-IPM is based on the assumption that the motion of the individual electrons can be described classically and there is no interaction between the electrons during the short collision times involved. The failure of the model is attributed mainly to its single particle character, i.e., the neglect of the electron–electron interaction.

In the nCTMC approach (n-body CTMC), the classical Hamiltonian is solved for $N+2$ particles (N electrons plus 2 nuclei), constituting $6(N+2)$ coupled, first-order differential Newton-type equations of motion

$$H = \frac{P_p^2}{2M_p} + \frac{P_T^2}{2M_T} + \sum_{i=1}^{N} \frac{p_i^2}{2m_i} + \sum_{i=1}^{N} \frac{Z_T Z_i}{R_{Ti}} + \sum_{i=1}^{N} \frac{Z_p Z_i}{R_{pi}} + \frac{Z_p Z_T}{R}, \quad (2.45)$$

where the indices p and T refer to the projectile and target, respectively. Electron–electron correlation neglected in nCTMC model can be extremely

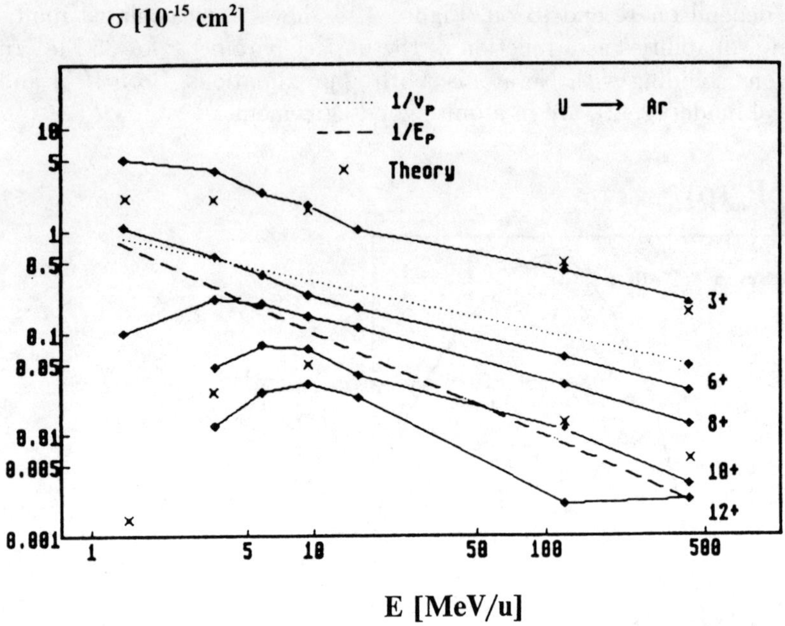

Fig. 2.15. Multiple ionization cross sections for U^{q+} ions on Ar. The charge states $q = 44$, 55, 65, 65, 75, 90 and 91 for energies of 1.4, 3.9, 5.9, 9.4, 15.5, 120 and 420 MeV/u, respectively (experiment). Charge states of recoil ions are indicated. The crosses are nCTMC calculations for recoil charge states of 3+ and 10+ [2.135]

important for light systems (e.g., H^+ + He collisions) where the accurate representation of the double-ionization cross sections requires the direct inclusion of electron–electron interactions (see, e.g., [2.44, 62, 131]). The electrons initially start in a microcanonical distribution about the target nucleus with the effective Z for each electron determined by the sequential binding energies of the target atom. The net-ionization cross section can be described by CTMC which highly overestimates the experimental values for high m.

The CTMC and the Russek models overestimate cross sections for high charge states m. CTMC also fails to account properly for the production of high charge states but provides the results that agree with experiment for the total cross sections and singly and doubly-differential cross sections.

Statistical Model. A statistical model (or Energy Deposition model, ED) of multielectron processes in collisions of atoms with *slow* ions was developed by *Russek* et al. [2.132–134]. This model assumes two steps in the collision. In the first step, the projectile transfers its kinetic energy to the excitation of the target electrons. After the collision (the second step), this excitation

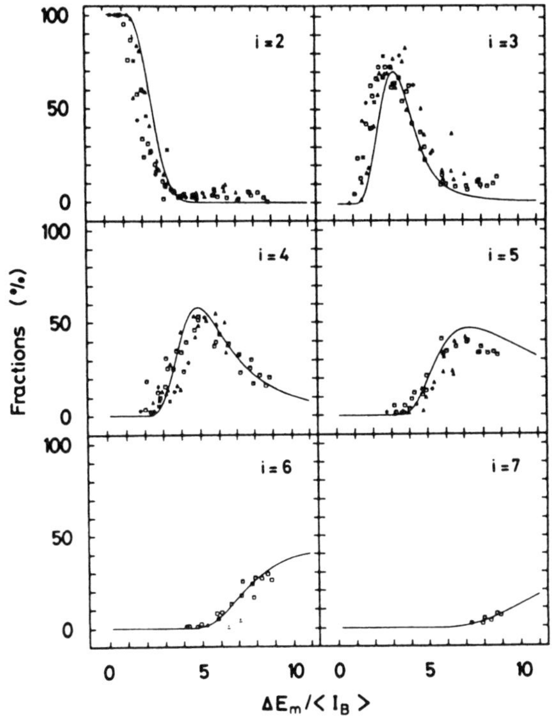

Fig. 2.16. Charge-state fractions of target ions B^{m+} produced in reaction (2.46) with $k = 2$ as a function of the reduced energy $\Delta E_m / < I_B >$: symbols – experiment, solid lines – statistical model. From [2.136]

energy is distributed among electrons leading, under certain conditions, to the target multiple ionization. The model is based on the assumption that the probability for the formation of a collision induced final state with m electrons in the continuum is obtained assuming that its value is proportional to the volume of phase-space available at that ionization state. The probability depends on the energy transferred by the projectile (deposited energy) which is assumed to be statistically distributed among all electrons in the system.

The Russek model was successfully used for statistical interpretation of transfer ionization in slow collisions of multiply charged ions with atoms [2.136] at keV energies:

$$A^{q+} + B \to A^{(q-k)+} + B^{m+} + (m-k)e^-, \ k = 2 \text{ or } 3, \tag{2.46}$$

with projectiles A = C, N, Ne, Ar, Kr and Xe and targets B = Ne, Ar, Kr and Xe. Figure 2.16 shows charge-state fractions of target ions B^{m+} produced in reaction (2.46) with $k = 2$ as a function of the reduced energy $\Delta E_m / < I_B >$

where $< I_B >$ is the averaged ionization potential of the target and ΔE_m is the maximum potential energy given by ionization energies I of the A^{j+} and B^{j+} particles:

$$\Delta E_m = \sum_{j=q-k}^{q-1} I_A^{(j)} - \sum_{j=0}^{k-1} I_B^{(j)} . \quad (2.47)$$

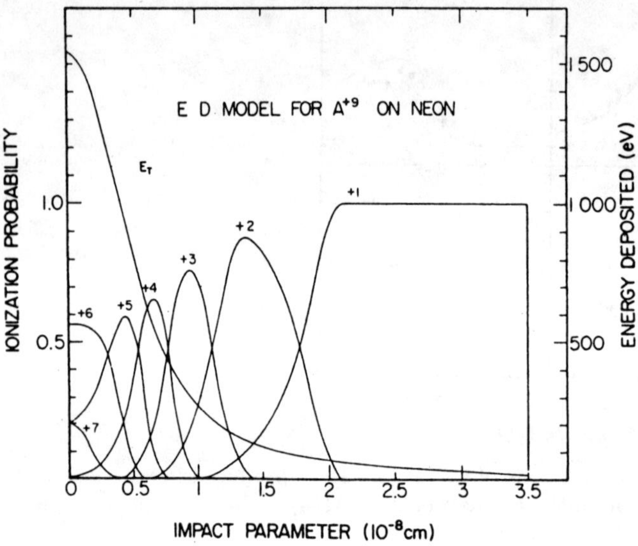

Fig. 2.17. The calculated energy E_T deposited by 1 MeV/u 9-fold charged ions colliding with Ne and multiple ionization probabilities P_m, $m = 1$–7, as a function of the impact parameter. From [2.104]

The ED model was extended by *Cocke* [2.104] to the region of *fast* ion–atom collisions. He also suggested to consider the energy deposition as being roughly due to the fast passage of a point charge particle through a cloud of electrons. The latter was considered as a gas of free classical electrons. However, his model overestimated the cross sections for low stages of ionization which was attributed to the roughness of calculation of the energy deposition. Figure 2.17 shows calculated by the ED model, the calculated energy deposited with in 1 MeV/u A^{9+} ions colliding with Ne and the esponding probabilities P_i for the i-fold ionization. Multiple ionization cross sections in collisions of 34 MeV Cl^{q+} ions with Ne and Ar are shown in Fig. 2.18 together with experimental data.

Semiclassical model. This time-dependent (TD) model was developed by *Horbatsch* and co-workers [2.137–139] to solve the time-evolution of N-

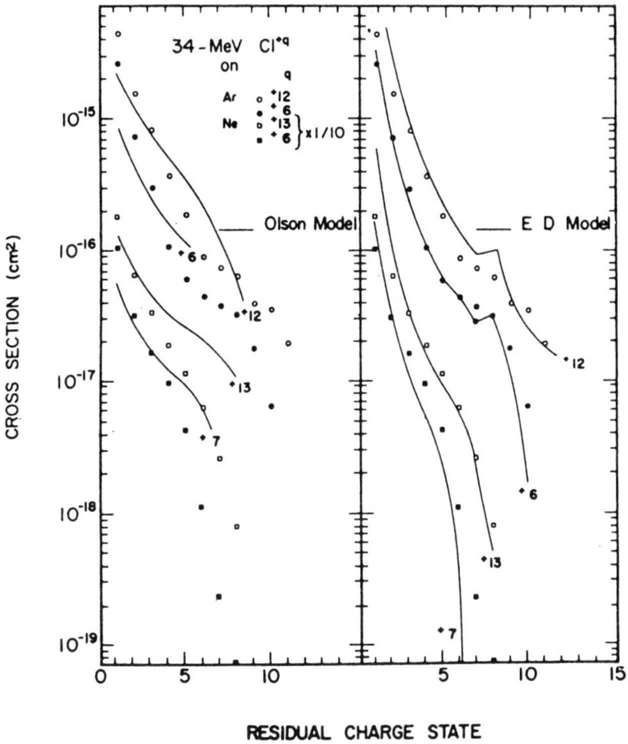

Fig. 2.18. Cross sections for the production of recoil charge states q for 34-MeV Cl in selected charge-states of Ar. Olson model – CTMC model [2.103], ED – energy deposite model; symbols – experimental data. From [2.104]

electron system for a given nuclear trajectory by means of a single distribution function. The theory deals not with the quantum mechanical description but with classical or semiclassical ones. The problem of N electrons and two colliding nuclei is usually reduced to an explicitly time-dependent N-body problem with a fixed nuclear trajectory. Then TD many-body problem is reduced to IPM, i.e., TDHF to TDTF or Vlasov model. The Vlasov equation is solved by the CTMC method: solving Newton's equations for randomly selected test particles. To calculate the population in the charge state of the target the binomial distribution is used.

A model is quantum-statistical in nature, based on the description of time-dependent many particle systems in terms of quantum statistical model, the so-called semiempirical quantum statistical mean-field approach. The time evolution of the many-electron density is calculated from the classical Vlasov equation after the initial distribution in a phase space and a model Hamiltonian have been specified. The model is statistical in nature

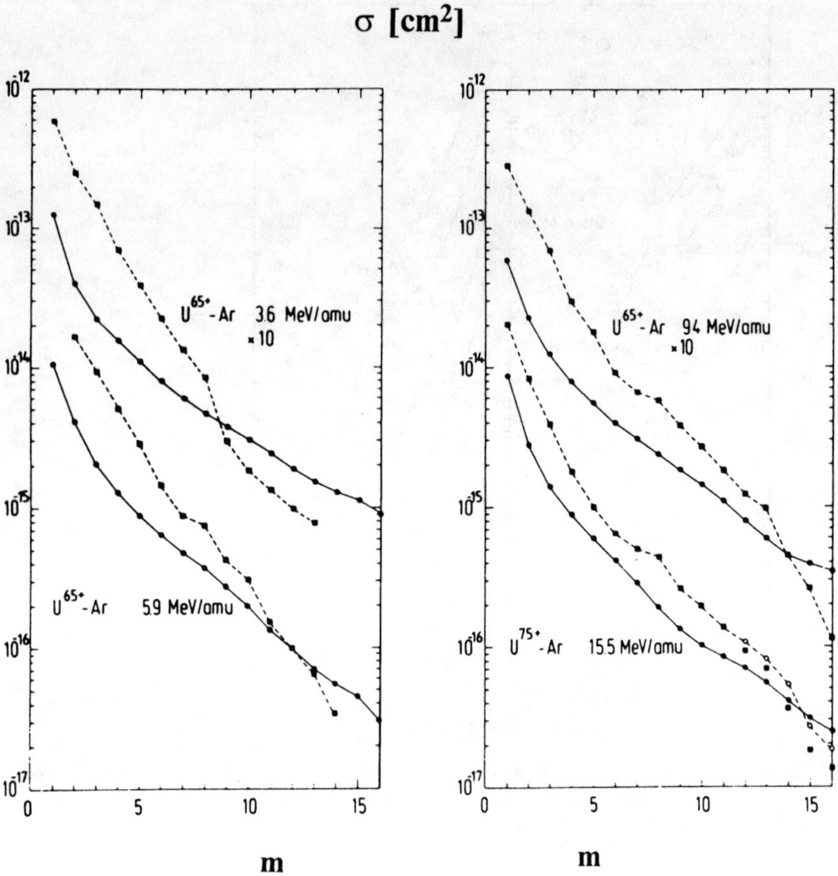

Fig. 2.19. Production cross sections of Ar^{m+} recoil ions as a function of m in $U^{65+,75+}$ - Ar collisions at 3.6–15.5 MeV/u energies: dashed curves – experimental data for pure ionization [2.70,107], solid curves – semiclassical theory. From [2.140]

(similar to TDHF approach) and thus averages over any shell structure, and is based on the solution of the Vlasov equation with a certain effective potential:

$$\frac{\partial}{\partial t} f(\mathbf{r}, \mathbf{p}, t) + \nabla_{\mathbf{p}} H \cdot \nabla_{\mathbf{r}} f(\mathbf{r}, \mathbf{p}, t) - \nabla_{\mathbf{r}} H \cdot \nabla_{\mathbf{p}} f(\mathbf{r}, \mathbf{p}, t) = 0. \tag{2.48}$$

In a single particle model, the Hamiltonian H is given by

$$H(\mathbf{r}, \mathbf{p}, t) = \frac{p^2}{2} - \frac{Z_{\text{eff}}(t)}{r} - \frac{q}{|\,\mathbf{r} - \mathbf{R}(t)\,|} \tag{2.49}$$

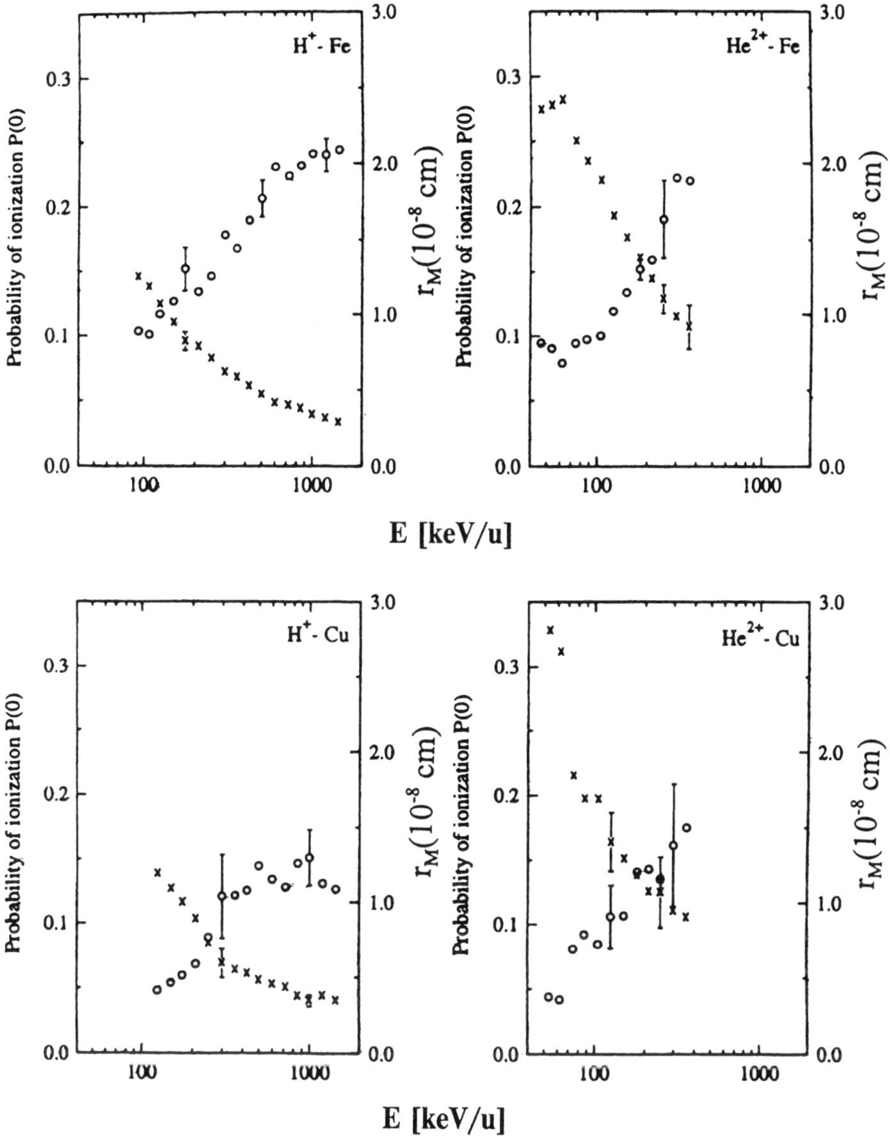

Fig. 2.20. Single-electron probabilities $P_s(0, v)$ and fitting parameters [r_M in (2.51)], obtained from experiment. A few representative limits of uncertainties are shown. From [2.109]

with q the projectile charge and $Z_{\text{eff}}(t)$ the effective nuclear target charge. In (2.48), $f(\mathbf{r}, \mathbf{p}, \mathrm{t})$ is the atomic distribution function (the single particle density operator). The function f is calculated with the initial condition:

$$f(\mathbf{r}, \mathbf{p}, t = 0) = N \frac{\mathrm{e}^{-ar}}{(p^2 + a^2/4)^4}, \tag{2.50}$$

where N is the number of electrons and a is the screening parameter adjusted to give the statistical atom a proper static potential. The parameter $a \approx 3.0$ for Ne, Ar, Kr and Xe targets (see [2.140] for detail). This theory overestimates the direct ionization and underestimates charge-exchange cross sections. To predict the target with a very high charge, m varies orders of magnitude with a slight variation of Z_{eff}. A typical example for production of Ar^{m+} recoil ions in collisions of uranium ions with argon atoms is depicted in Fig. 2.19.

The time evolution of the electron-charge cloud as a function of the nuclear motion for a fixed impact parameter is in many features similar to what one would obtain in a time- dependent HF calculations.

Semiempirical Approach. The IPM model was applied to describe the pure ionization processes in collisions of positive ions with Ar, Fe, Cu and Ga targets [2.94, 109, 112].

A simple method for fitting multiple ionization cross sections by the single probability P_s for a given shell was developed by *DuBois* and *Manson* [2.101] assuming P_s in the form:

$$P_s(\rho, v) = P_s(0, v) \exp(-\rho/r_M), \tag{2.51}$$

where $P_s(0, v)$ and r_M are the fitting parameters. The form, (2.51), has been proved for large impact parameters. The single ionization probability for a given shell independent of ρ and obtained by fitting the intergated ionization cross section, (2.43), with experimental data for low m-values is displayed in Fig. 2.20. The measured and calculated multielectron ionization cross sections in collisions of protons and α-particles with Cu atoms are shown in Fig. 2.21.

This procedure is quite effective to describe the pure ionization processes caused by direct ionization of the outermost electrons, i.e., for low m-values. For higher charge-state recoil ions, this semiempirical formula underestimates the experimental data which is most probably due to the contribution from the inner-shell ionization followed by Auger electron emission and vacancy cascade resulting in the enhancement of higher charge ions (Fig. 2.22).

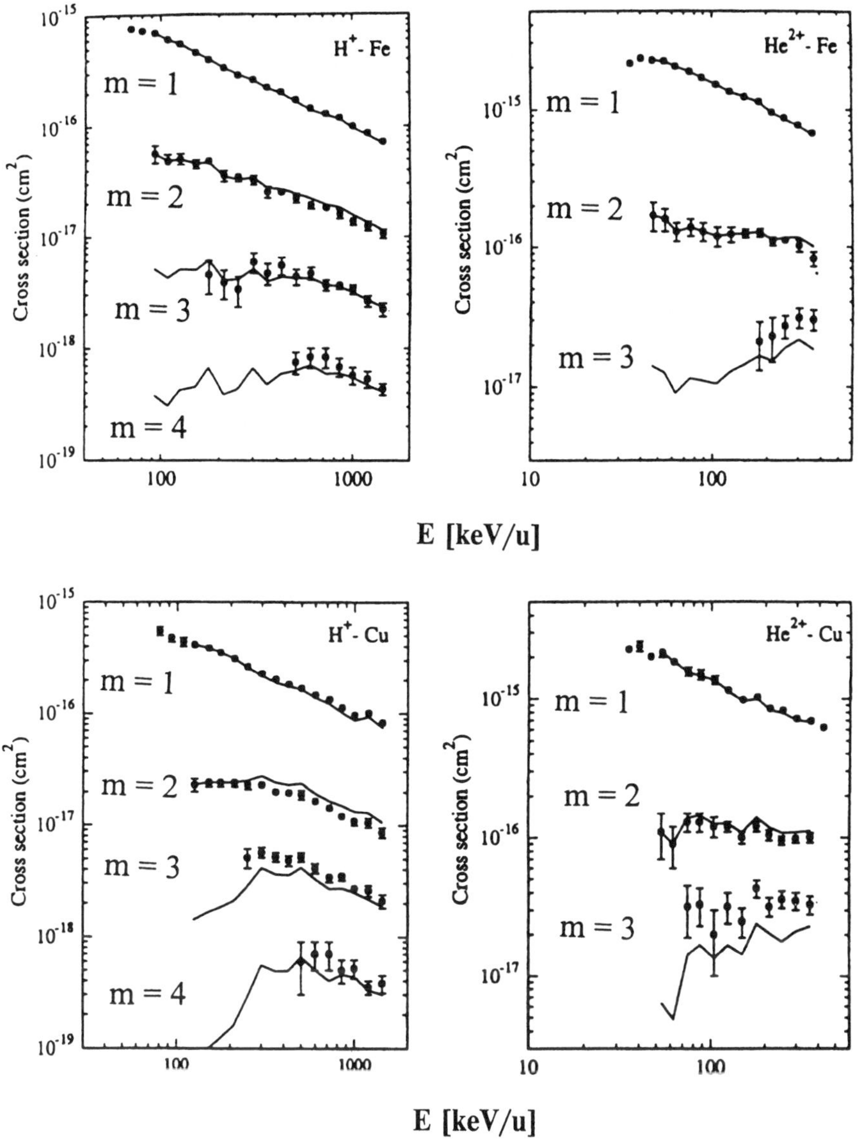

Fig. 2.21. Pure m-electron ionization cross sections in collisions of H^+ and He^{2+} with Fe and Cu: symbols – experiment, solid curves – fitting by (2.43, 51). From [2.109]

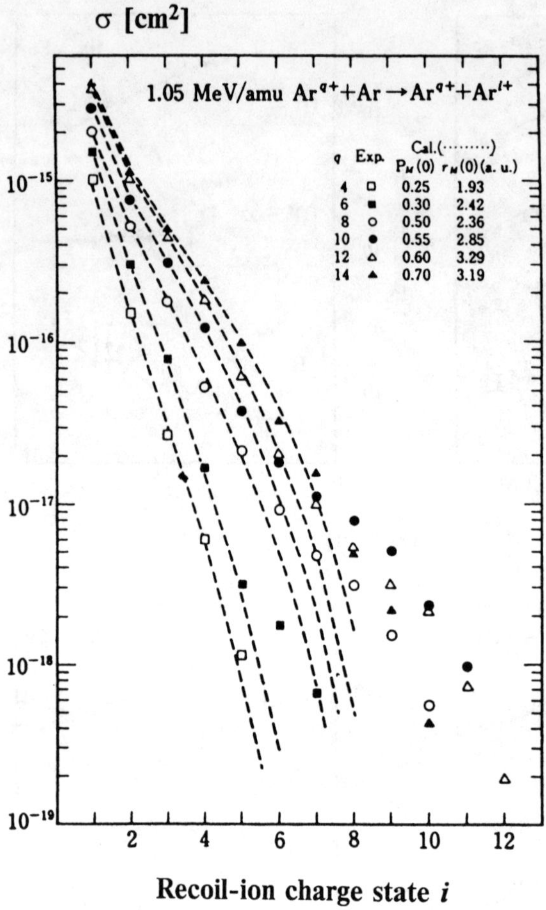

Fig. 2.22. Pure ionization cross sections for 1.05-Mev/u Ar^{q+} on Ar target in comparison with a two-parameter fit, (2.51), for the M-shell of Ar atom. For recoil ion charge states $m > 6$, experimental data (symbols) are higher than theoretical estimate because of the possible contribution of the indirect processes related with ionization of the L-shell electrons. From [2.94]

Scaling Laws. The CTMC calculations showed [2.98] that the calculated values of the net-ionization cross section σ_+ for a given rare-gas target can be described by the scaling relation

$$\tilde{\sigma} = \sigma_+/q, \quad \tilde{E} = E/q\,[\mathrm{keV/u}]\,, \tag{2.52}$$

where q is the projectile charge, $\tilde{\sigma}$ and $\tilde{E}$ are scaled cross section and energy, respectively. It gives quite good results for many projectiles and spicies in the low- and intermediate-energy ranges (Fig. 2.23). At high energies

CTMC calculations fail because, based on purely classical methods, this approach does not comprise the logarithmic energy term $(\ln E)/E$ which is much stronger than the classical term E^{-1}, especially for highly-charged projectiles.

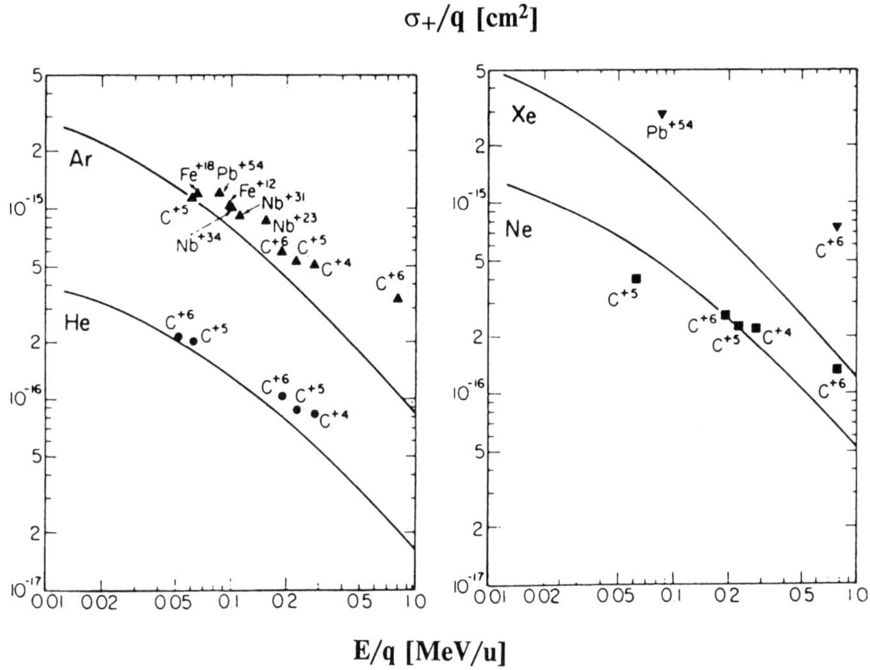

Fig. 2.23. Scaled net-ionization cross sections in collisions of highly charged ions with rare-gas atoms: symbols – experiment, solid curves – CTMC calculations. From [2.98]

The q^2-scaling for the multiple ionization cross sections on the projectile charge q

$$\tilde{\sigma} = \sigma_m(E)/q^2, \quad \tilde{E} = E\,[\mathrm{keV/u}] \tag{2.53}$$

was applied for the ionization cross sections of Fe, Cu and Ga targets by light projectile, H^+ and He^{2+} [2.109] (Figs. 2.13, 24). This scaling is based on the first Born approximation [2.5].

Another scaling for multiple ionization cross sections in collisions of highly charged ions with atoms was suggested in [2.141]

$$\tilde{\sigma} = \sigma_m(E)/q^{0.7}, \quad \tilde{E} = E/q^{1.6}\,[\mathrm{keV/u}] \tag{2.54}$$

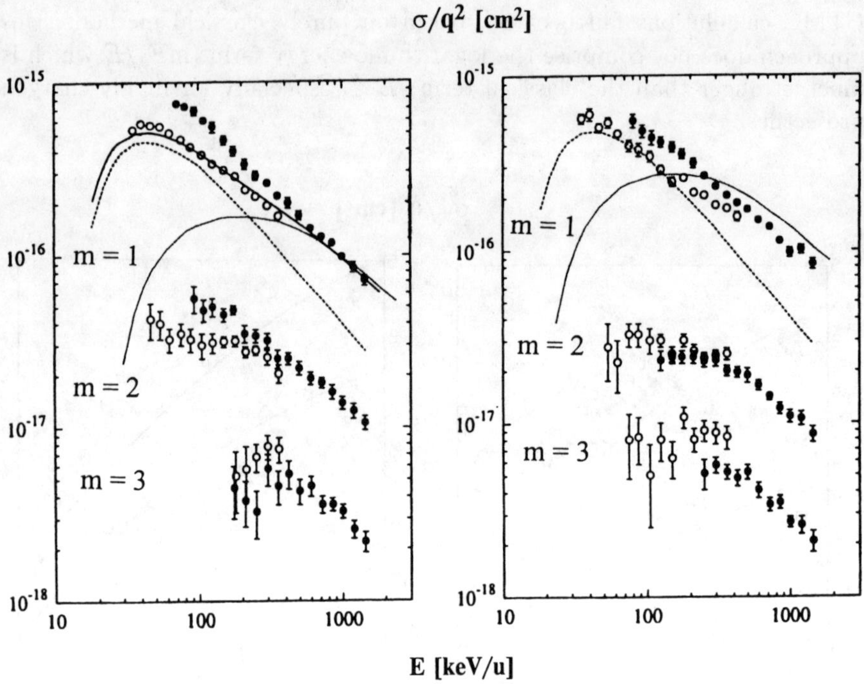

Fig. 2.24. Scaled multiple ionization cross sections in collisions of H^+ and He^{2+} with Cu and Fe: experiment – symbols, theory (1st order approximation): solid curves – total single ionization, dashed curves – ionization of a $4s$-electron, dotted curves – ionization of a $3d$-electron. From [2.109]

The results for ionization of Ar atoms by protons and highly charged ions are shown in Fig. 2.25.

In general, there is no adopted scaling for the pure multiple ionization cross sections so far because the experimental data are quite scarce and also there is no adequate theory for the description of these processes.

2.3 Multiple Excitation

From the theoretical viewpoint, multielectron excitation processes caused by electron impact and heavy charged particles are more understandable than multiple ionization where the continuum Coulomb wave functions are not known. However, measurements of multiple-excitation cross sections seem to be much more difficult compared to those of multiple ionization. Most of studies have been performed involving double excitation of atoms and ions from the ground ns^2-configuration (He, Be, Ca, H^-, O^{4+}) to the lowest-lying

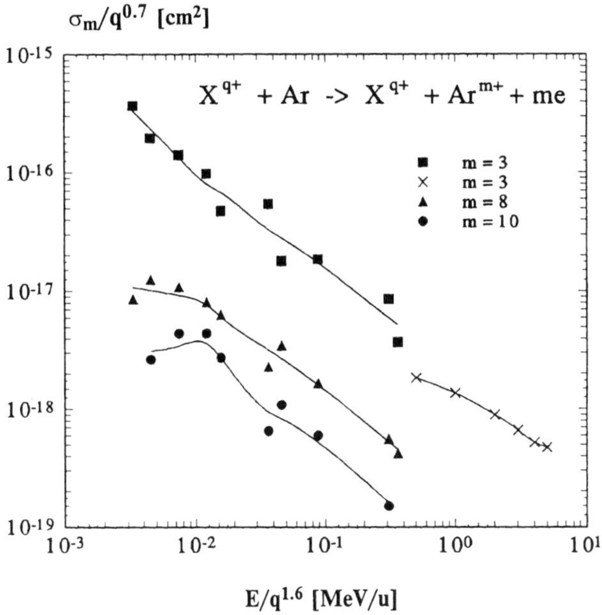

Fig. 2.25. Scaled m-electron ionization cross sections of Ar by highly charged ions (q >40) and protons (q = 1, crosses): symbols – experimental data from [2.44, 106, 107, 135]; solid lines are to quide the eye. From [2.141]

doubly excited states. Investigations of two-electron transitions provide a deep insight into dynamics of atomic collisions and, especially, the nature of electron–electron correlation effects in many-electron phenomena [2.5, 142].

2.3.1 Double Excitation by Electrons

Theory and experiment for double-electron excitation cross sections $\sigma^{**}(E)$ are mainly devoted to studying the atomic systems with two valence s-electrons, i.e., to transitions $s^2 \to n_1l_1n_2l_2$. Measurements of the relative $\sigma_2(E)$-values for He atoms have been carried out in [2.143–145] for transitions $1s^2 \to 2s2p, 2s^2$ and $2p^2$. Double excitation of He constitutes a special problem because the doubly excited states of He all lie above the first ionization limit and decay mostly via autoionization.

The simplest model used for description of many-electron processes is the independent-particle model (IPM, Sect. 2.2) which assumes that each target electron moves independently each other in the average atomic core potential. Deviations from the IPM predictions are usually attributed to the electron–electron correlation effects which indicate that the motions of all target electrons are directly related.

Theoretical investigations of double excitation cross sections are reported in [2.146–150] using the perturbative approaches and in [2.151–156] using the close-coupling method. The experimental electron emission yields and theoretical predictions for double-electron transitions in He induced by electrons, protons and highly charged ions are given in Table 2.6.

Table 2.6. Experimental emission yields and theoretical (close-coupling approach) double-excitation cross sections (in 10^{-20} cm^2) of He at 1.5 MeV/u due to electrons, protons and highly charged ions. The results for the ^1D and ^1P are summed in the last three columns in order to compare with experimental and theoretical data. From [2.145]

	$2s^2(^1S)$		$2p^2(^1D)$		$2s2p(^1P)$		$2p^2(^1D)$ and		$2s2p(^1P)$
Projectile	Exp. [2.145]	Th. [2.151]	Exp. [2.145]	Th. [2.151]	Exp. [2.145]	Th. [2.151]	Exp. [2.145]	Exp. [2.144]	Th. [2.151]
e^-	0.0816	0.73		0.27	1.17	3.3	3.05	1.92	3.57
p	0.0318	0.74	1.84	0.48	0.608	3.0	2.45	2.75	3.48
C^{4+}	4.56	15.6	63.8	38.0	8.32	60.0	72.1	47.2	98.0
C^{5+}	8.08		104		20.6		124.6	100.4	
C^{6+}	10.9	51.6	170	156	31.1	162	201.1	149.9	318.0
F^{7+}	16.8		218		22.8		241		
F^{8+}	45.4		299		40.7		340		
F^{9+}	62.9	198	456	603	33.9	513	489		1116

In the first-order approximation, the matrix element $< \phi_0(\mathbf{r_1}, \mathbf{r_2}) \mid V \mid \phi_1(\mathbf{r_1}, \mathbf{r_2}) >$ of two-electron transition equals zero [2.146, 157] if the atomic wave functions $\phi_{0,1}$ of the initial 0 and final 1 states are orthogonal; here V is the interaction of the incident electron with two optical electrons. If the transition takes place via an intermediate state, for example, $s^2 \to sp \to p^2$, then $\sigma^{**} \neq 0$, and is given in the second-order perturbation approximation by

$$\sigma^{**} = \frac{8\pi a_0^2}{g_0 E} \int_{k_0-k_1}^{k_o+k_1} K dK \sum_{M_0, M_1} \mid A(\mathbf{K}) \mid^2, \quad \mathbf{K} = \mathbf{k_0} - \mathbf{k_1}, \tag{2.55}$$

where K is the momentum transfer, k is the incident electron momentum, M is the projection of the atomic orbital momentum L and g_0 is the statistical weight. Usually, transitions to the intermediate states are optically allowed transitions. The contribution from the second-order amplitude due to transition via an intermediate state a is given by [2.146]:

$$A^{(II)}(\mathbf{K}) = \int d\mathbf{R} e^{i\mathbf{KR}} \left[d\mathbf{x} G_a(x) V_{1a}(\mathbf{R} - \mathbf{x})\, e^{i\mathbf{k_1 x}} \right] V_{a0}(\mathbf{R}), \tag{2.56}$$

where $G_a(x) = \exp(ik_a x)/(2\pi x)$ is the Green function of the free motion, and $V_{ab}(x)$ are transition matrix elements.

Double-excitation cross section can be nonzero already in the first-order approximation if the correlation between atomic wave functions is taken into account. In this case one has

$$\sigma^{(I)} = \frac{8\pi a_0^2}{g_0 E} \int_{k_0-k_1}^{k_o+k_1} \frac{dK}{K^3} \sum_{M_0,M_1} |< \phi_0 \mid \mathrm{e}^{\mathrm{i}\mathbf{K}\mathbf{r}_1} + \mathrm{e}^{\mathrm{i}\mathbf{K}\mathbf{r}_2} \mid \phi_1 >|^2, \tag{2.57}$$

with

$$\begin{aligned} &< \phi_0(r_1, r_2) \mid \phi_1(r_1, r_2) > \\ &\quad = \lambda_0[< \tilde{\varphi}_0(r) \mid \varphi_1(r) >]^2 + \lambda_1[< \varphi_0(r) \mid \tilde{\varphi}_1(r) >]^2, \end{aligned} \tag{2.58}$$

where $\lambda_{0,1}$ are the mixture coefficients and $\varphi(r)$ and $\tilde{\varphi}(r)$ are the s-electron wave functions (see [2.147] for detail).

Double-excitation cross sections have different asymptotic dependence on the incident electron energy E and the *target* ion charge Z_T:

$$\sigma^{(I)} \sim Z_T^{-4} E^{-1}, \quad \sigma^{(II)} \sim Z_T^{-6} E^{-2}, \quad E \to \infty. \tag{2.59}$$

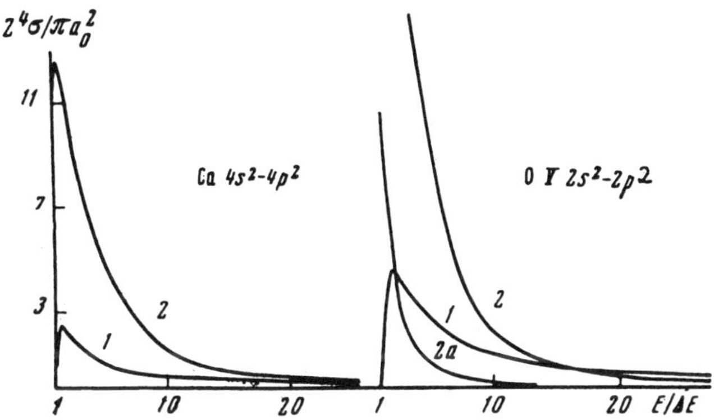

Fig. 2.26. Double-excitation cross sections for transitions Ca($4s^2 - 4p^2$) and OV($2s^2 - 2p^2$) as a function of scaled electron energy $E/\Delta E$ (ΔE is the double-electron transition energy): 1 – first-order perturbation theory, (2.57), 2 – second-order perturbation theory, (2.55), 2a – curve 2 divided by a factor of 10. From [2.147]

Calculated double-excitation cross sections for transitions in Ca($4s^2 - 4p^2$) and OV($2s^2 - 2p^2$) are displayed in Fig. 2.26. As seen from Fig. 2.26, at small electron energies E the cross section is given by the second-order approximation and the correlation effects are not significant. At high energies, the opposite situation takes place: the first-order approximation plays a key role when it is necessary to account for the configuration interaction.

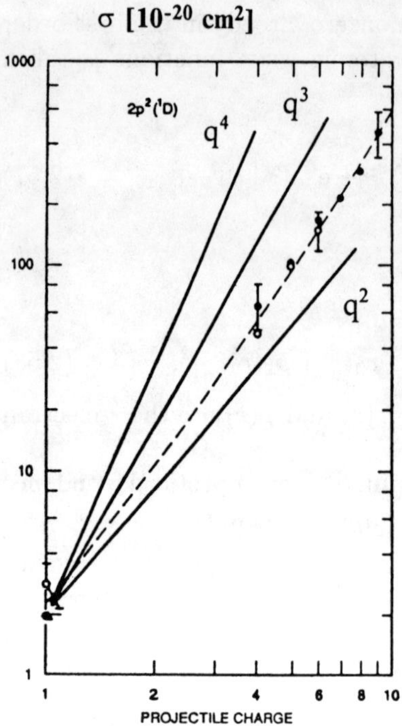

Fig. 2.27. The total averaged emission cross sections from the $2p^2\,{}^1D$ state as a function of the projectile charge q at energy of 1.5 MeV/u. Electrons are indicated by triangles. Open symbols are experimental data from [2.144]. The dashed line is a fit to $\sigma^{**}(q) = 1.92 \times 10^{-20}\,\mathrm{cm}^2\,q^a, a = 2.47 \pm 0.43$. The solid lines indicate the q^2, q^3 and q^4 dependences. From [2.145]

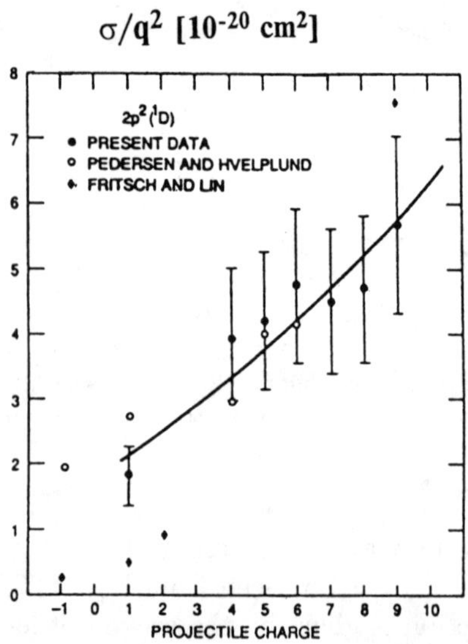

Fig. 2.28. Scaled total averaged emission cross sections from the $2p^2\,{}^1D$ state as a function of the projectile charge q at ion energy of 1.5 MeV/u. The open circles are the data from [2.144], the diamonds represent the calculations [2.151]. The solid curve is the fit to (2.60) and does not imply that the cross section for a neutral projectile is not zero. From [2.145]

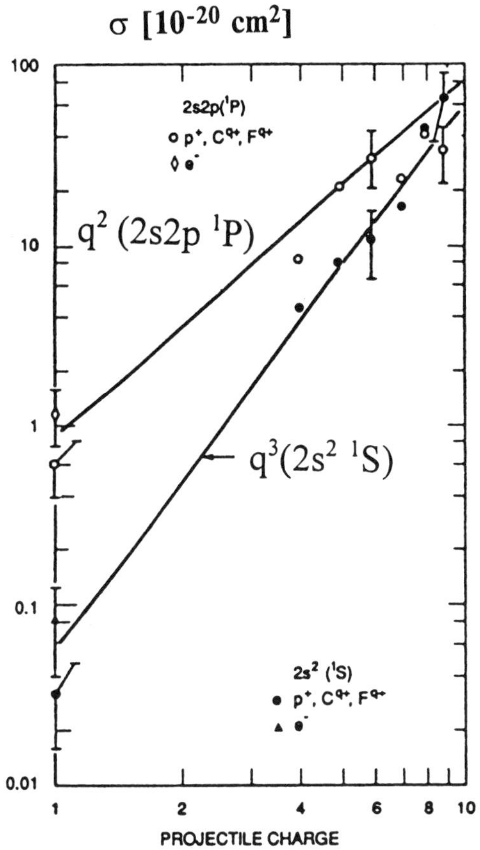

Fig. 2.29. The total averaged emission cross sections from the $2s^2\,{}^1S$ and $2s2p\,{}^1P$ states as a function of the projectile charge q. From [2.145]

For ions ($Z_T > 1$), the cofiguration interaction is larger than for neutral targets. From threshold up to the region of the cross section maximum, the σ^{**}-values can be described by the close-coupling method.

2.3.2 Double Excitation by Positive Ions. Projectile-Charge Dependence

Cross sections for double excitation are also quite scarce and reported for He, Be, H^-, Ar^{16+} and some other systems [2.144, 145, 158, 159]. Some experimental and theoretical data for excitation of He atoms into lower-lying states are presented in Table 2.6. The data show different dependence on the *projectile* charge q: for transitions $1s^2 \to 2s^2({}^1S)$ and $2p^2({}^1D)$ the cross section σ^{**} increases approximately as q^3 while for excitation into $2s2p({}^1P)$ state it varies approximately as q^2. These dependences, reproduced also in Figs. 2.27–2.29, are much less than q^4-dependence followed from the IPM.

They show the importance of the interaction between two target electrons in creating the doubly excited states. Therefore, the experimental data are usually described by a parametrization formula

$$\sigma^{**}(q) = a_1 q^2 + a_2 q^3 + a_3 q^4, \tag{2.60}$$

where the coefficient a_2 is proportional to the strength of the quantum interference between the first- and second-order mechanisms [2.145]. The first-order mechanism can be attributed to the single transitions via intermediate states while the second-order mechanism can be modeled as two successive single excitations. Theoretical aspects of the problem are discussed also in [2.150, 154].

2.4 Simultaneous Ionization and Excitation

Although the ionization processes with simultaneous excitation of the target are more complicated as compared to the double excitation, the experimental data on the cross sections σ^{+*} for the first double-electron process are presented in many papers mostly for He atoms colliding with fast electrons, protons and light ions [2.160–165] and also for Ne atoms [2.161] and H_2 molecules [2.166]. Ionization with excitation processes can strongly influence the intensities of the Lyman series of the He^+ ions [2.167, 168] and other applications in laboratory and astrophysical plasmas. Theories of ionization with excitation processes have been developed mainly for high-energy collisions using the first-order perturbation approximation or its modifications [2.169–174], the close-coupling expansion [2.175] or an R-matrix method [2.176].

Experimental cross sections [2.165] for ionization-excitation of He atoms by electrons and protons from the ground into excited np-states

$$\mathrm{e}^-, \mathrm{H}^+ + \mathrm{He}(1s^2\,{}^1S_0) \to \mathrm{e}^-, \mathrm{H}^+ + \mathrm{He}^{+*}(np\,{}^2P^o) + \mathrm{e}^-, \quad n = 2-5, \tag{2.61}$$

are displayed in Fig. 2.30. The most striking feature of these figures is that the cross sections for electrons are approximately twice as large as for protons at relative velocities $v > 6$ a.u. *Bailey* et al. [2.165] attributed this to the quantum interference between first- and second-order amplitudes $\sim (Z_p/v)^3$ but not to the mass effect.

Similar results for He^+ resonance line $2p - 1s$ induced by proton and molecular hydrogen ion impact are displayed in Fig. 2.31. There is a good agreement between measurements of different groups except for the region of about $v = 1.4$ a.u. *Bailey* et al. [2.165] explained this difference by a possible Coulomb explosion of the weakly bound molecular projectiles colliding with the target gas. Although the total production of screened fragments is small

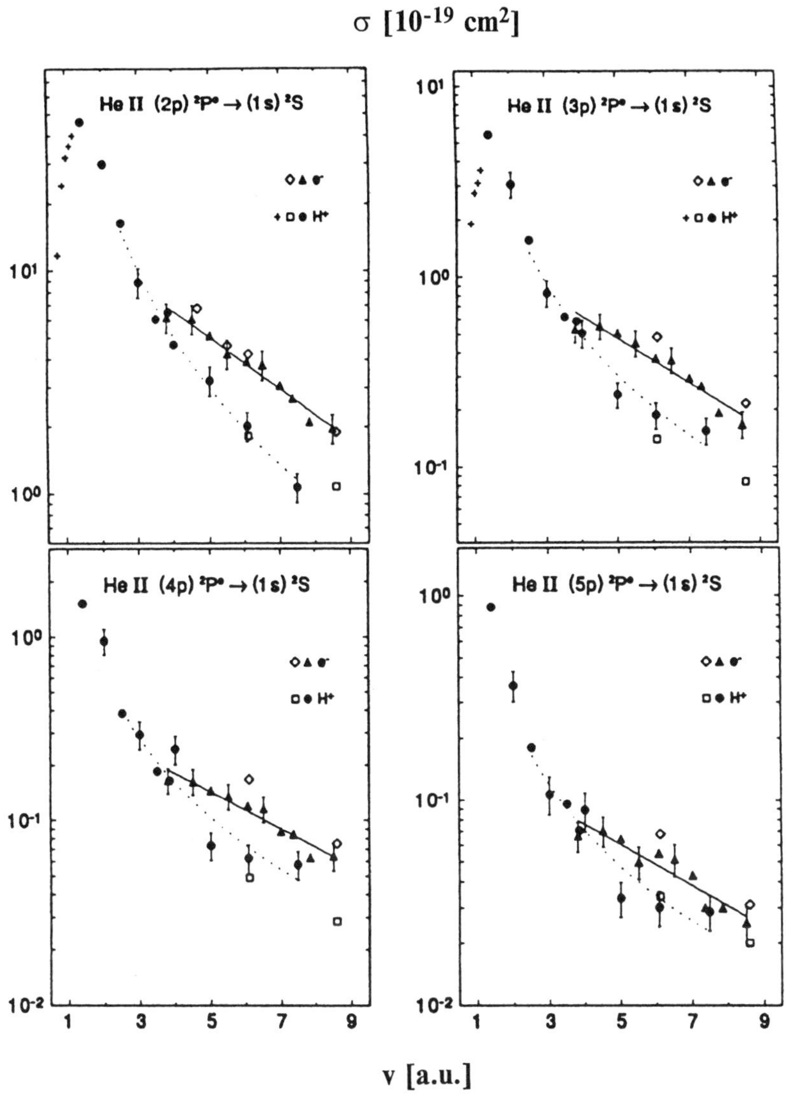

Fig. 2.30. Experimental cross sections for ionization-excitation of helium for the transitions $He^+(np)\,^2P^0 \to (1s)\,^2S, n = 2 - 5$: solid triangles, e^-, solid circles, H^+, [2.165]; diamonds, e^-, squares, H^+, [2.163]; crosses, H^+, [2.160]. Solid curves are to quide the eye. From [2.165]

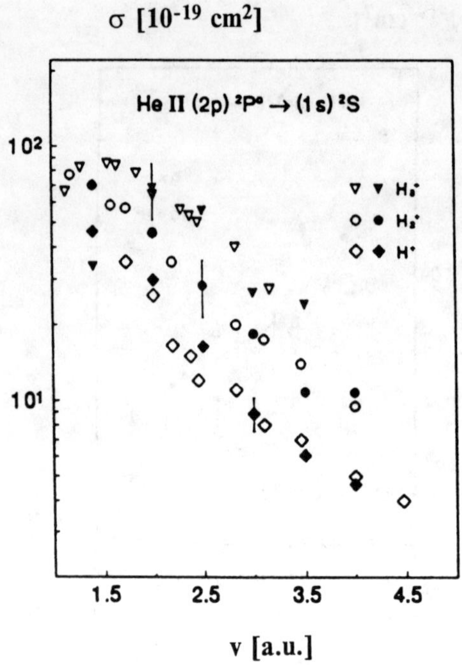

Fig. 2.31. Experimental cross sections for ionization-excitation of helium to the $He^{+*}(2p)\,^2P^o)$ level in H_3^+, H_2^+ and H^+ impact. Full symbols – [2.165], open symbols – [2.177, 178]. From [2.165]

at the present energies, predissociation processes of the projectile prior to entering the emission region of the target cell may account for this difference, indicating that each heavy particle (proton) behaves independently.

The first experimental results on the state-selective cross section ratios $\sigma^{+*}(np)/\sigma^*(1snp)$, $n = 2$–4, in He for electron and proton impact in the $v = 3.5$–10 a.u. velocity range were reported in [2.164]. The results displayed in Fig. 2.32 show also a factor of 2 difference between electron and proton impact cross sections at high impact energies (opposite signs of the projectiles) that reflects the importance of dynamical electron–electron interactions in two-electron transitions. *Fülling* et al. [2.164] have demonstrated that the ratio $\sigma^{+*}(np)/\sigma^*(1snp)$ approaches the double-to-single ionization ratio σ_{2+}/σ_{+} in the limit of high principal quantum numbers $n \to \infty$. In the case of electrons as projectiles, the trend of the ratio σ^{+*}/σ^* is shifted towards the lower energies when compared to the σ_{2+}/σ_{+} ratio is probably due to the smaller momentum transfer involved.

Calculations of ionization-excitation cross sections of He atoms by electrons from the ground $1s^2$ and excited $1s2s\ ^{1,3}S$ states performed within an R-matrix method are shown in Figs. 2.33, 2.34. For the final state $He^+(2p)$ (Fig. 2.33), the R-matrix calculations are in a quite good agreement with the close-coupling results [2.175] but only in a qualitative agreement with ex-

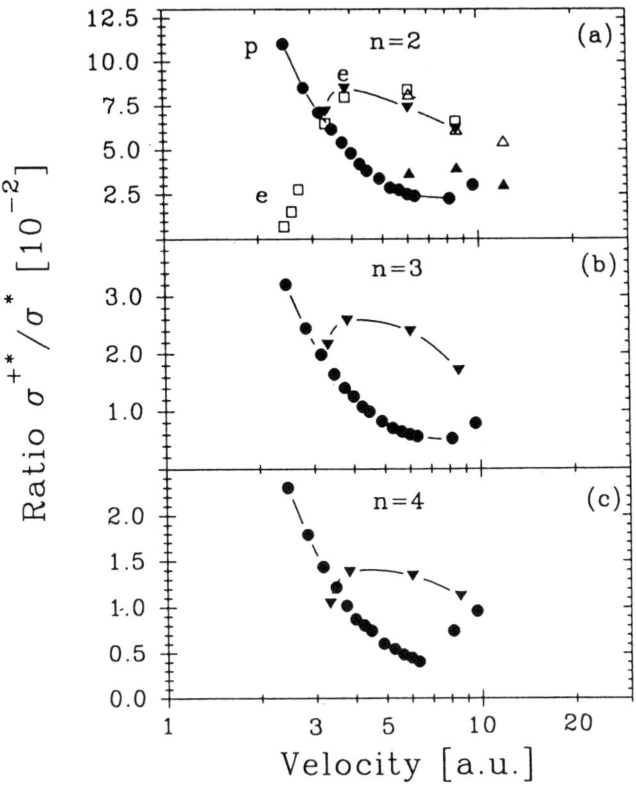

Fig. 2.32. Experimental cross section ratios $\sigma^{+*}(np)/\sigma^{*}(1snp), n = 2-4$: solid circles, protons, solid triangles (down), electrons – [2.164]; squares, electrons – [2.167]; solid triangles (up), protons and open triangles, electrons – [2.163]. From [2.164]

perimental data. In the resulting $He^{+}(2s)$ ion, the close-coupling data [2.175] are approximately two times larger than R-matrix results.

In the case of ionization-excitation from the excited state $He^{*}(1s2s\ ^1S)$, calculations performed by the R-matrix method are in agreement with the Born-type calculations [2.170] although in the light of the comparison for the total cross section this may be fortuitous. Unfortunately, no experimental data exsist to check the theoretical calculations.

According to [2.5], a large difference between ionization-excitaion cross sections of He induced by electron and proton impact is expected because of the quantum interference between first- and second-order perturbation theory contribution. Therefore, there are two basic reaction mechanisms contributing to the observed cross sections: few-electron processes where the projectile interacts only once with the target and the second-order processes

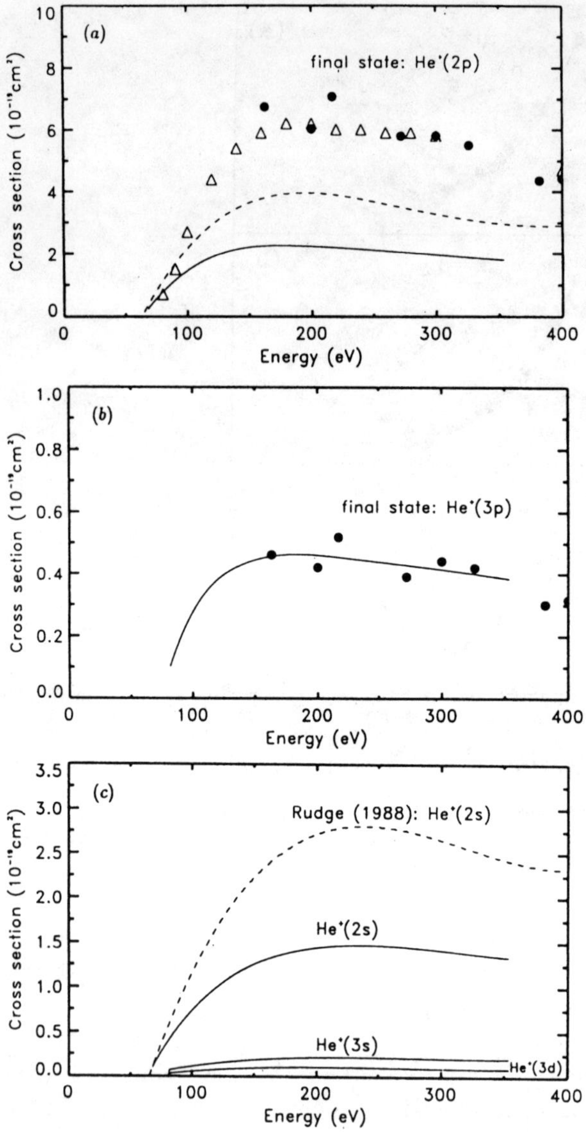

Fig. 2.33. (a) Cross sections for electron-impact ionization of He($1s^2\ ^1S$) with simultaneous excitation into $2p\ ^2P$ state of He^+ as a function of the incident electron energy. Theory: solid curve – an R-matrix calculations [2.176]; dashed curve – close-coupling calculations [2.175]. Experiment: solid circles – [2.164], open triangles – [2.167]. (b) Same as (a) for $He^+(3p\ ^2P)$. (c) Same as (a) for $2s\ ^2S, 3s\ ^2S$ and $3d\,^2D$ final states. From [2.176]

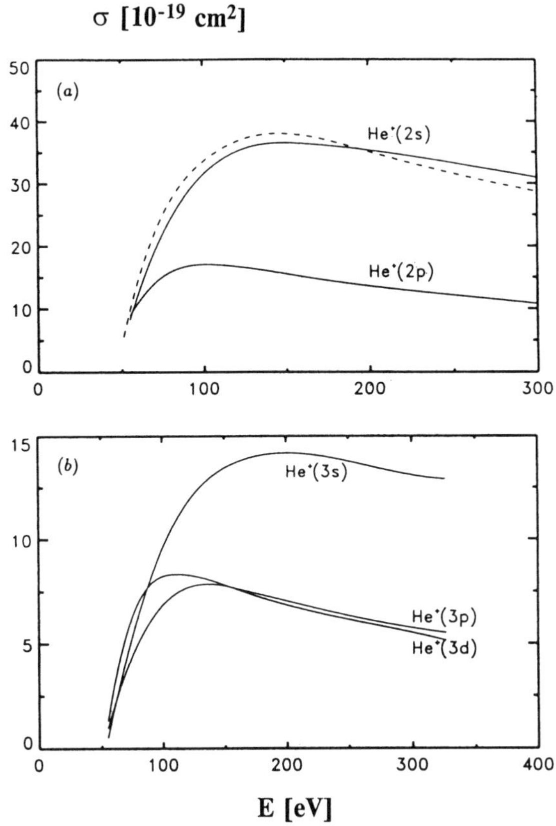

Fig. 2.34. (a) Cross sections for electron-impact ionization of $He^*(1s2s\ ^1S)$ with simultaneous excitation into $2s\ ^2S$ and $2p\,^2P$ states of He^+ as a function of the incident electron energy. Theory: solid curve – an R-matrix calculations [2.176]; dashed curve – close-coupling calculations [2.170]. (b) Same as (a) for $3s\ ^2S, 3p\ ^2S$ and $3d\,^2D$ final states. From [2.176]

where the projectile interacts twice with the target in a single collision. The difference in behavior is explained by the presence of $(Z_p/v)^3$ term since the cross section is proportional to the square of the amplitude integrated over the impact parameter:

$$\sigma^{+*} = 2\pi \int \mid A \mid^2 \rho d\rho = 2\pi \int \mid A^I + A^{II} \mid^2 \rho d\rho, \tag{2.62}$$

where A^I is the first-order transition amplitude proportional to Z_p/v and A^{II} is the second-order amplitude proportional to $(Z_p/v)^2$.

In the work [2.172], the two-electron cross sections for e^-+He (Z_p= -1) and H^+ + He (Z_p= +1) collisions is described by a semiempirical formula

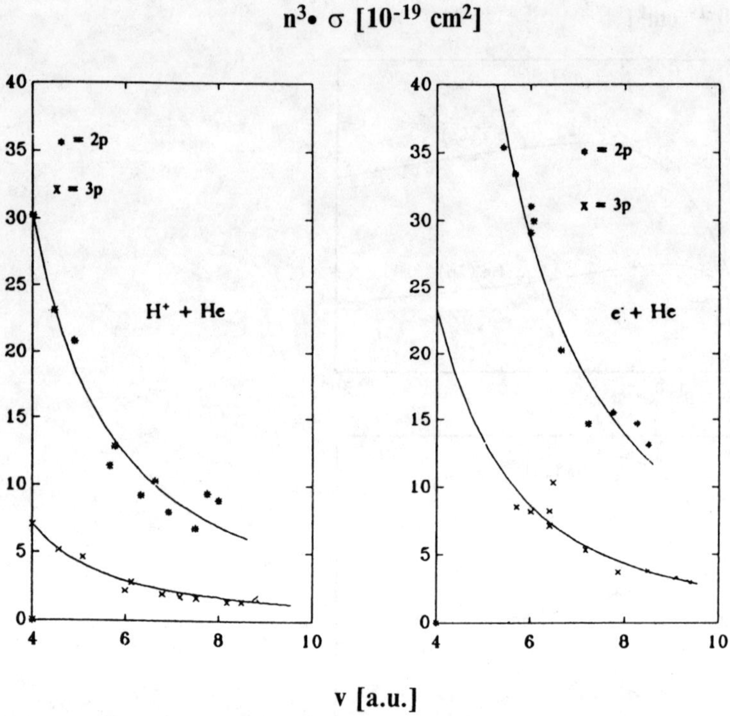

Fig. 2.35. Scaled ionization-excitation cross sections of He to $n = 2, 3$ levels by electrons and protons: symbols – experiment [2.164], solid curves – the fit, (2.63), and Table 2.7. From [2.172]

$$\sigma^{+*}(v) = 10^{-19}\,\mathrm{cm}^2 \cdot \sum_{i=0}^{2} C_{i,n}\,(Z_p\,v_0/v)^i, \tag{2.63}$$

where v_0 is the atomic velocity unit ($v_0 = 2.2 \times 10^8$ cm/s) and $C_{i,n}$ are the fitting parameters obtained by a least-square method from experimental data [2.164]. The coefficients $C_{i,n}$ are given in Table 2.7 for electron and proton impact as a function of the principal quantum number n for the $He^+(np)$ Rydberg states. The fit of this parametrization with experimental data is plotted in Fig. 2.35. As seen from Fig. 2.35, the fit is quite satisfactory for protons and electrons, meaning that the first-order, second-order and cross-terms are sufficient to describe experimental data at velocities considered. The third-order contributions are expected to be much smaller [2.154].

Table 2.7. Expansion coefficients $C_{i,n} \times 10^3$ for electron and proton impact ionization-excitation cross sections into the excited of $He^+(np)$ states, $n = 2$–5, (2.63). From [2.172]

		i	
n	0	1	2
2	0.580	1.768	5.573
3	0.128	0.520	1.850
4	0.051	0.306	1.936
5	0.055	0.234	1.145

2.5 Multielectron Capture

Multielectron capture processes

$$A^{q+} + B \rightarrow A^{(q-m)+} + B^{m+}, \quad m \geq 2, \tag{2.64}$$

arising in collisions between positive ions and neutral atoms, besides a pure theoretical interest, are of importance for several fields of applications such as controlled nuclear fusion research, plasma physics, physics of energetic heavy ions in the Earth's magnetosphere and others. The contribution from multielectron capture processes can reach up to 25–30% and, therefore, these processes should be included in the calculations of the beam attenuation kinetics. In contrast to one-electron capture where a lot of data were measured and succesfully described by several theories, the transfer of $m \geq 2$ electrons constitutes a much more complicated problem although an increasing number of experimental cross sections have come up recently for the capture of up to 6 electrons. At present, there is no theory existing which could describe accurately the multiple-electron capture processes.

Multielectron capture as well as other atomic multielectron processes are governed by different interaction mechanisms at low, intermediate and high energies. Therefore, in these energy ranges, the multielectron capture cross sections are described by different dependences of the projectile charge, the nuclear charge of the target, the number of the captured electrons and relative velocity. Consequently, the scaling laws will be different for these energies.

2.5.1 Very Low Energies

Cross sections for electron transfer in the very low-energy region, E = (0.01–100)q eV/u, were reported in [2.179–181, 184–192] applying different techniques including ion spectroscopy, Auger electron spectroscopy, photon spectroscopy, and coincidence technique.

A large amount of experimental cross sections obtained for collisions of positive ions having the charge q as high as 8 with the gas targets at low velocities ($v \ll 1$ a.u.)

$$A^{q+} + B \to A^{(q-m)+} + B^{m+}, \quad 2 \leq q \leq 8 \tag{2.65}$$

$$A = \mathrm{Ne, Ar, Kr, Xe},$$
$$B = \mathrm{He, Ne, Ar, Kr, Xe, H_2, N_2, O_2, CH_4, CO_2} \tag{2.66}$$

are reported in [2.179–182]. These experimental material allow us to obtain the semiempirical scaling law for relatively low q-ions and for m-electron capture cross sections $\sigma_{q,q-m}$ in the form

$$\sigma_{q,q-m} = 10^{-12}\mathrm{cm}^2 \cdot C(m) q^{a(m)} (I_B/\mathrm{eV})^{-b(m)}, \; v \ll 1 a.u., \tag{2.67}$$

where I_B is the first ionization potential of the target, C, a and b are the fitting parameters given in Table 2.8. Some of the experimental multielectron

Table 2.8. Fitting parameters for m-electron capture cross sections at low velocities, (2.67). From [2.181]

m	$C(m)$	$a(m)$	$b(m)$
1	1.43±0.76	1.17±0.09	2.76±0.19
2	1.08±0.95	0.71±0.14	2.80±0.32
3	(5.50±5.8)×10^{-2}	2.10±0.24	2.89±0.39
4	(3.57±8.9)×10^{-4}	4.20±0.79	3.03±0.86

capture cross sections are displayed in Fig. 2.36 vs the projectile charge q for Ar^{q+} + Xe collisions at 30 keV projectile energy.

A similar scaling law, based on the classical over-barrier model [2.210, 211], for multiple electron capture involving relatively high-charge ions has been proposed [2.183].

An example of the capture cross sections systematically measured at very low energies $E = (0.125 - 25)q$ eV/u for Ar^{q+} + Ne collisions ($q = 7-9$) is shown in Fig. 2.37. For single-electron capture cross sections (SC), experimental data are compared with the semiempirical formula (2.67), the absorbing sphere model [2.193] and with the Langevin cross section:

$$\sigma_L = \pi q \left(2\beta/E_{cm}\right)^{1/2}, \tag{2.68}$$

where β is the polarizability of the target atom and E_{cm} is the collision energy in the center-of-mass system [2.194].

It is seen that while single-electron capture cross section remains to be nearly constant in the whole energy range considered, the multiple-changing cross sections have a strong energy dependence with a minimum.

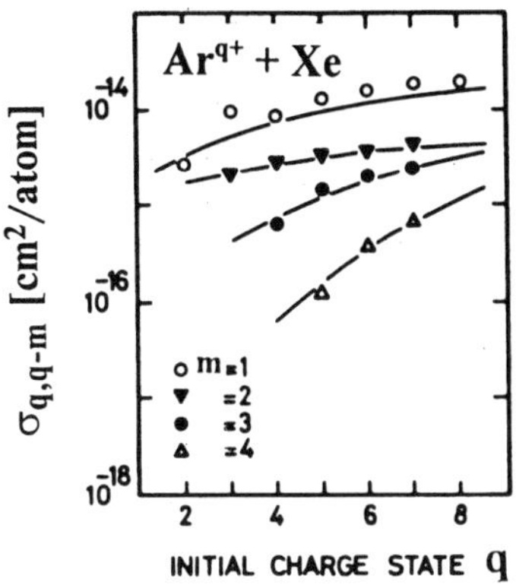

Fig. 2.36. The m-electron capture cross sections for 30 keV Ar^{q+} ions incident on Xe: symbols – experiment, curves – (2.67). From [2.181]

It should be noted that in some collision systems involving molecules such as $He^{2+} + H_2$ at very low energies, the double-electron capture becomes far dominant (almost two orders of magnitude) over the single-electron capture [2.195]. This can be understood from the fact that strong repulsive potential curves of doubly ionized molecular ions can find some accidental resonance levels in projectile ions.

2.5.2 Low and Intermediate Energies

Experimental data on electron-capture cross sections at moderately high velocities (E = 10–1000 keV/u) were reported [2.196–205], mainly for double-electron capture of rare-gas atoms by bare ion projectile (H^+ to Al^{13+}) in the ground state. (However, to our knowledge, there have been no measurements published with the determination of the initial projectile-ion population). The status in the field of the double-electron capture was presented in [2.206, 207].

For theoretical description of multiple-electron capture processes in this energy range, the classical over barrier model introduced by *Bohr* and *Lindhardt* [2.209] and then developed by *Bárány* et al. [2.210] and *Niehaus* [2.211] is used which gives the simpliest picture to understand multielectron capture and to estimate the cross sections.

Numerical calculations of the multiple-electron capture cross sections have been performed only for double-electron capture using the close-coupling method [2.212–214], the IPM model [2.121], or a quasimolecular treatment [2.215].

Figure 2.39 displays the behavior of low-energy double-electron capture cross sections into particular excited states in reaction

$$N^{7+} + He(1s^2) \to N^{5+**}(3d^2) + He^{2+}, \qquad (2.69)$$

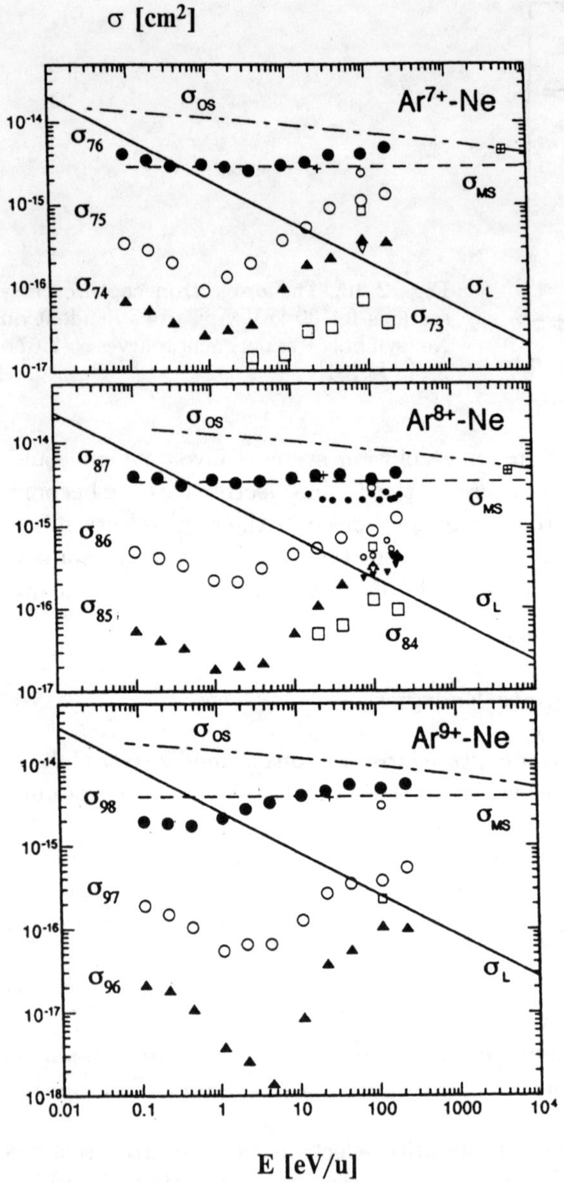

Fig. 2.37. The m-electron capture cross sections in collisions of Ar^{q+} (q = 7–9) with Ne. Experiment: large solid circles, single-electron capture (SC), large open circles, double capture (DC), large solid triangles up – triple capture (TC), large open squares – quadruple capture (QC) – [2.192]; middle open circles, SC, middle open squares, circles, DC, open triangles down, TC – [2.186]; small solid circles, SC, small open circles, DC, solid triangles down, TC – [2.187]; crosses, SC – [2.188]; crossed squares, SC – [2.184]. Theory: dashed curves (MS) – scaling law (2.67) for SC; dot-dashed curves – absorbing sphere (OS) model [2.193]; solid curves – Langevin model, (2.68). From [2.192]

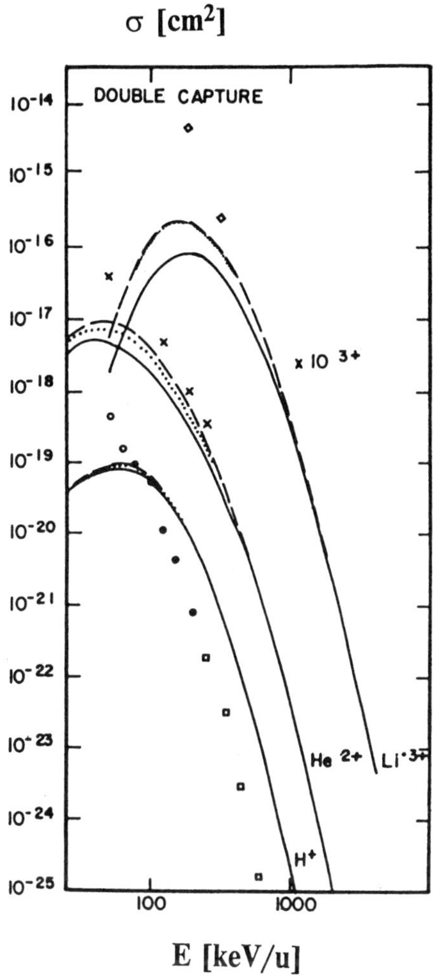

Fig. 2.38. Cross sections for double-electron capture to the ground state in the charge transfer of the H^+, He^{2+} and Li^{3+} nuclei in He vs nuclei energy. Experiment: open circles – H^+ [2.196], solid circles – H^+ [2.200], squares – H^+ [2.198], crosses – He^{2+} [2.199], diamonds – Li^{3+} [2.197]. Theory: curves – IPM calculations with different normalization of the single-capture probability (see [2.121] in details). From [2.121]

where experimental data are compared with the calculations [2.215] using two-electron quasimolecular diabatic states with the use of multichannel models. The best agreement with experiment is obtained if all nonadiabatic regions of the quasimolecular terms are taken into calculations.

It is well known that in slow highly charged ions colliding with multielectron targets, many electrons can be transfered into projectile ions with high probabilities ($\sim 10^{-16}$ cm^2), resulting in the formation of the high Rydberg states which, in turn, decay mostly via autoionization processes. In 4 keV/u Xe^{27+} + Xe collision, the recoil secondary ions with the charge state as high as Xe^{16+} which are observed in coincidence with the projectile Xe^{22+} ions.

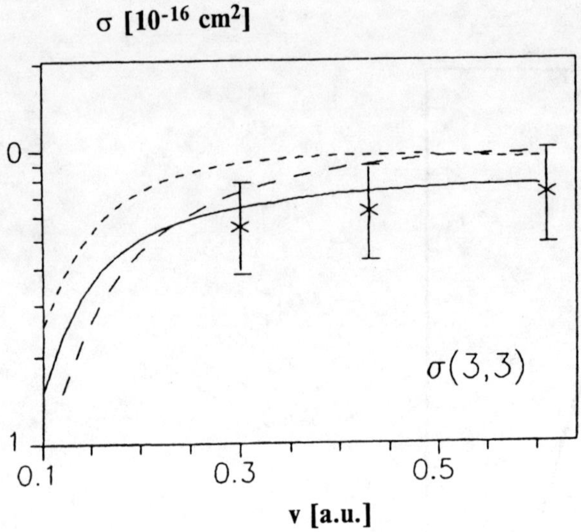

Fig. 2.39. Double-electron capture cross sections into $N^{5+}(3d^2)$ state in N^{7+} + He($1s^2$) collisions. Experiment: crosses – [2.216], theory: solid curve – with the account of all nonadiabatic regions (NR), long-dashed curve – only one NR at internuclear distance R = 5.1 a.u. was considered, short-dashed curve – with account of one NR and configuration interaction of only two configurations. From [2.215]

This means that more than 10 electrons among 16 electrons captured are autoionized! (see [2.208]).

One of the important problems arising in the double-electron capture is the population of the excited states over the quantum numbers n and ℓ of two electrons captured into the projectile. There are two distinguished cases of the resulting ion $A^*(n\ell, n'\ell')$: symmetrical (or equivalent electron) configurations with $n \sim n'$ or asymmetrical (or nonequivalent electron) configurations with $n >> n'$. Symmetrical configurations are populated mainly at relatively high colliding energies, while asymmetrical at lower ones. In the energy range of several ten keV, these features are devoted to two population mechanisms *monoelectronic* and *dielectronic* considered in [2.201]. Mechanisms for double-electron capture in slow ($v = 0.04$–1.0 a.u., E = 0.5–250 keV) Ne^{10+} + He collisions have been studied in [2.204] using the high-resolution Auger spectroscopy (Fig. 2.40). The double-electron capture cross sections to the $3\ell n\ell'$ ($n \geq 6$) states increase strongly when the collision energy decreases from 250 keV to 10 keV (0.15 a.u.) and become dominant at sub-keV energies.

In Fig. 2.38 the double-electron capture cross sections are presented for H^+, He^{2+} and Li^{3+} ions colliding with He atoms. Calculations have been

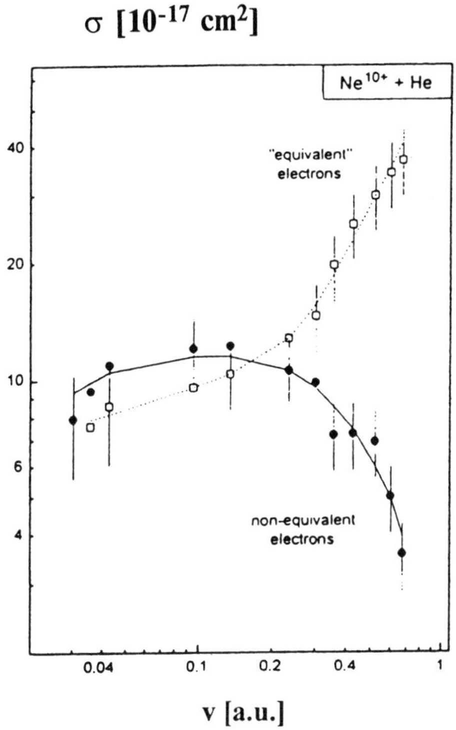

Fig. 2.40. Experimental total cross sections for producing $n\ell n'\ell'$ states in Ne^{10+} + He collisions as a function of relative velocity. The squares correspond to the equivalent-electron configurations $3\ell n\ell', n = 4$–5 and $4\ell n\ell', n = 4$–6, while the circles are associated with the nonequivalent electron configurations $3\ell n\ell', n \geq 6$. From [2.204]

performed using the IPM model with the single probabilities normalized at low energies.

2.5.3 High Energies

Experimental data on multiple-electron capture of fast ($E > 1$ MeV/u) positive ions are quite scarce [2.217–222], especially for heavy highly-charged ions due to the difficulties of producing these ions in the high-velocity range. Some experimental data for many-electron capture cross sections are displayed in Figs. 2.41, 42 for O^{q+} + He, Ar collisions at MeV/u energies. Investigations of the double-electron capture of 4–12 MeV/u Ge^{31+} ions on Ne revealed that for small impact parameters and small impact energies, where the projectile velocity is nearly equal to the orbital velocity of the bound L-shell electrons in Ge, double-electron capture is a dominant capture process [2.222]. This is demonstrated in Fig. 2.43 where the capture probabilities calculated by the nCTMC model are displayed as a function of the impact parameter. A similar effect was observed in [2.223]. *Schlachter* et al. investigated 47 MeV-Ca^{17+} ions in close collisions with Ar atoms and

Table 2.9. Experimental double-electron capture cross sections (in cm^2/atom) in collisions of fast He^{2+} ions with gas targets as a function of He^{2+} energy. From [2.224]

E, MeV/u	N_2	Ne	Ar	Kr	Xe
0.126	1.83×10^{-17}	1.97×10^{-17}	5.46×10^{-18}		
0.336	3.21×10^{-19}	6.63×10^{-19}	1.13×10^{-19}		
0.392	1.97×10^{-19}			3.01×10^{-19}	1.11×10^{-18}
0.572	3.04×10^{-20}	3.99×10^{-20}	4.08×10^{-20}	1.71×10^{-19}	1.82×10^{-19}
0.718	7.66×10^{-21}	6.70×10^{-21}	1.50×10^{-20}		
0.726	9.19×10^{-21}	7.79×10^{-21}	1.83×10^{-20}	7.59×10^{-20}	4.61×10^{-20}
0.991	1.72×10^{-21}	1.13×10^{-21}	5.96×10^{-21}	2.20×10^{-20}	
1.01	1.78×10^{-21}	1.02×10^{-21}	5.12×10^{-21}	1.76×10^{-20}	4.47×10^{-21}
1.41	2.42×10^{-22}	2.09×10^{-22}	8.08×10^{-22}	2.45×10^{-21}	6.12×10^{-22}
1.79	4.91×10^{-23}	8.04×10^{-23}	2.02×10^{-22}	6.62×10^{-22}	2.55×10^{-22}
2.40			1.16×10^{-23}		

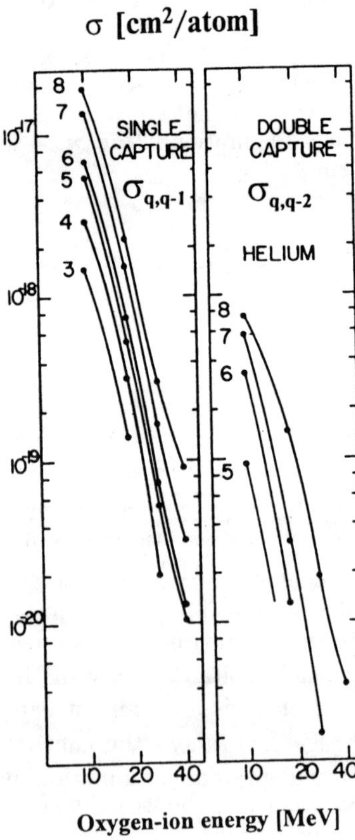

Fig. 2.41. Experimental electron-capture cross sections for oxygen ions in He. The smooth curves are labeled by the initial charge state. From [2.218]

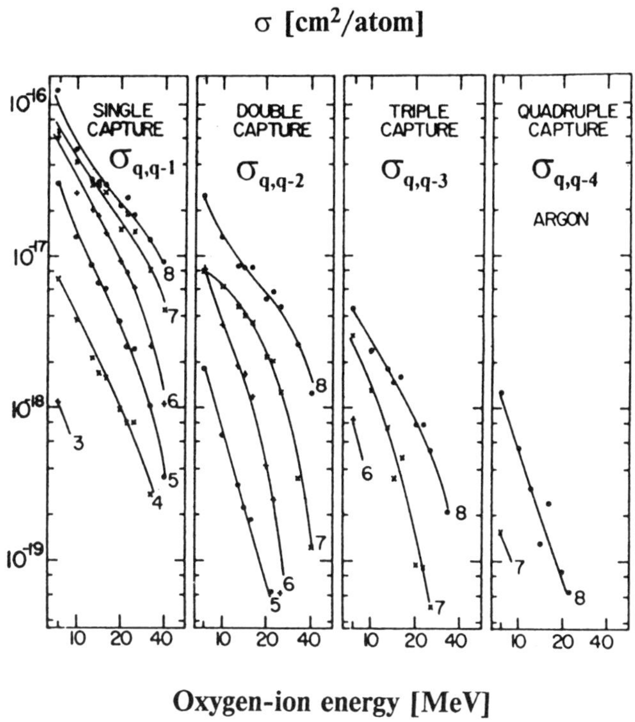

Fig. 2.42. Same as in Fig. 2.41 for Ar targets. From [2.218]

found a strong increase of the capture involving two and more electrons as compared to one-electron capture for close collisions. At higher energies (Fig. 2.43b), one-electron capture is again the most dominant capture process even at small impact parameters.

The measured absolute values of the double-electron capture cross sections from different gases by α-particles [2.224] are given in Table 2.9. The data have been meausred with the accuracy of $\pm 20\%$ using a technique of charge state analysis of fast particles after single collisions with the gas target. The data are in agreement with the previous measurements performed in [2.197, 225–228].

At farther high energy and relativistic energy region, the radiative electron capture (REC) process overcomes the non-radiative electron capture (NRC) process as the former becomes easier than the latter to get the total energy balanced. Though the single REC has been investigated relatively well (see [2.229]), the double REC process has not been confirmed.

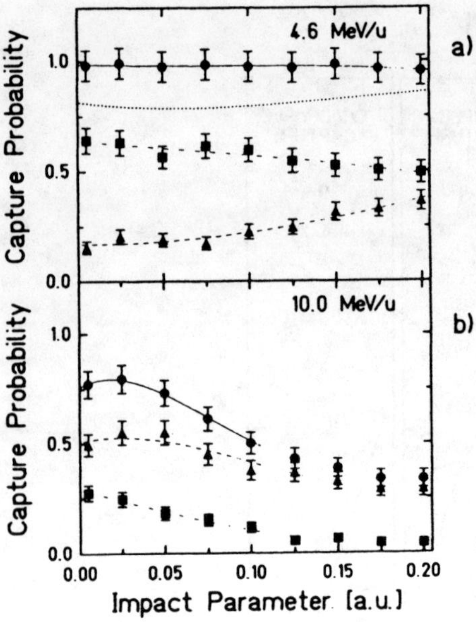

Fig. 2.43. Calculated impact-parameter dependence for capture of Ne electrons into Ge^{31+} ions according to nCTMC model at different energies (a) for 4.6 MeV/u and (b) for 10 MeV/u: circles and solid curves – total capture probability, triangles and dashed curves – single capture, squares and dot-dashed curves – double capture, dashed curves – the sum of single and double capture. From [2.222]

2.5.4 Scaling Law for Double-Electron Capture from He Targets

Scaling laws for multielectron capture have been considered. In [2.230], cross-section scaling for one- and two-electron loss processes in collisions of He with multicharged ions has been considered in a wide range of the projectile energies. For double-electron capture

$$A^{q+} + \mathrm{He} \to A^{(q-2)+} + \mathrm{He}^{2+}, \tag{2.70}$$

the following scalings have been proposed:

low-energy regime:

$$\tilde{\sigma}_{2c} = \sigma_{2c}/q = \mathrm{constant} = 1.37 \times 10^{-16}\mathrm{cm}^2, \quad E = 0.07 - 1.5\mathrm{keV/u}, \tag{2.71}$$

high-energy regime:

$$\tilde{\sigma}_{2c} = \sigma_{2c}/q = 5.8 \times 10^{-18}\mathrm{cm}^2/\tilde{E}^{-4.5}, \quad \tilde{E} = E[\mathrm{keV/u}]/(100q^{0.5})$$
$$E \geq 100q^{0.5}, \quad q \geq 6 \tag{2.72}$$

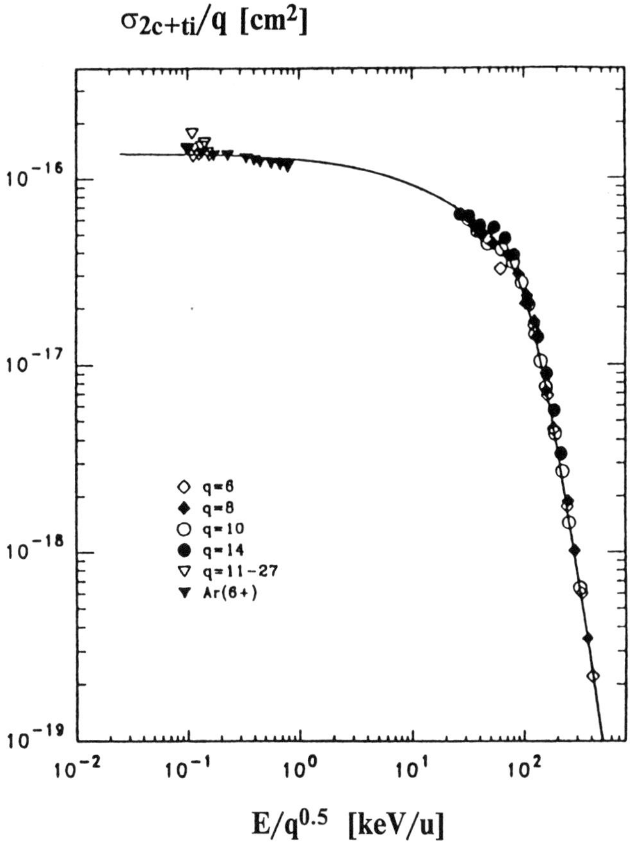

Fig. 2.44. Scaled cross section for a sum of double-electron capture and transfer ionization in collisions of He with multicharged ions as a function of reduced energy: symbols – experiment (see [2.230]), solid curve – fit, (4.19,2.75). From [2.230]

At energies $E > 80q^{0.5}$ keV/u, the two electron detachment from He atoms is dominated by the transfer ionization

$$\mathrm{A}^{q+} + \mathrm{He} \to \mathrm{A}^{(q-1)+} + \mathrm{He}^{2+} + \mathrm{e}^-, \tag{2.73}$$

the cross section σ_{ti} decreases more slowly than the double-electron capture cross section σ_{2c}. The total cross section, $\sigma_{2c} + \sigma_{ti}$, is represented in the form:

$$\tilde{\sigma}_{2c+ti} = \sigma_{2c+ti}/q = \frac{A\,B \times 10^{-16}[\mathrm{cm}^2]}{B + C\tilde{E}^a + \tilde{E}^4}, \quad \tilde{E} = E[\mathrm{keV/u}]/(100q^{0.5}), \tag{2.74}$$

$$A = 1.37, \; B = 0.64, \; C = 1.77, \; a = 0.75\,. \tag{2.75}$$

The scaled cross section for removal of two electrons from He atoms by highly charged ions is given in Fig. 2.44 together with experimental data. The fits (4.19) and (2.75) represent the experimental data to within 10 %.

3. Collisions Involving Negative Atomic Ions

This chapter treats the collision processes of negative atomic ions including antiprotons, namely, electron detachment or loss processes from negative ions under collisions with various particles as well as ionization and excitation of neutral atoms by negative ion impact. The electron detachment from negative ions can be theoretically treated as the electron loss or ionization into continuum which is dominant at high energies or as the electron transfer into projectile ions which is only important at low energies and decrease rapidly as the collision energy increases.

3.1 Electron Detachment from Negative Ions in Electron Collisions

The electron detachment from negative ions under electron impact has experimentally been investigated through so-called crossed-beams techniques which have been described in details in Chap. 1. Theoretical studies even on the single-electron detachment are limited and practically no theoretical analysis for multiple electron detachment has been reported. Only some empirical treatment of double-electron detachment processes are available and found to reproduce reasonably well the observed cross sections.

3.1.1 Negative Hydrogen Ions

The cross sections for the single- and double-electron detachment from the simplest negative ions, $H^-(1s, 1s')$, into continuum under electron impact

$$e^- + H^- \rightarrow e^- + H^0 + e^-, \tag{3.1}$$

$$e^- + H^- \rightarrow e^- + H^+ + 2e \tag{3.2}$$

have been measured (Fig. 3.1). The single-electron detachment process of the outer $1s'$ electron with the binding energy of 0.75 eV has been extensively investigated [3.1–3] and the observed cross sections seem to be in reasonable agreement with each other. The cross sections near the threshold region have been measured in detail [3.4, 5] using the H^- ion beam in a storage ring (Sect. 1.2.2). Some theoretical treatments have been reported, too [3.6, 7]. Recent calculations based upon the close-coupling theory, taking into account the Coulomb repulsive trajectory of the incident electrons, result in good agreement with the observed data even near the threshold energies [3.8].

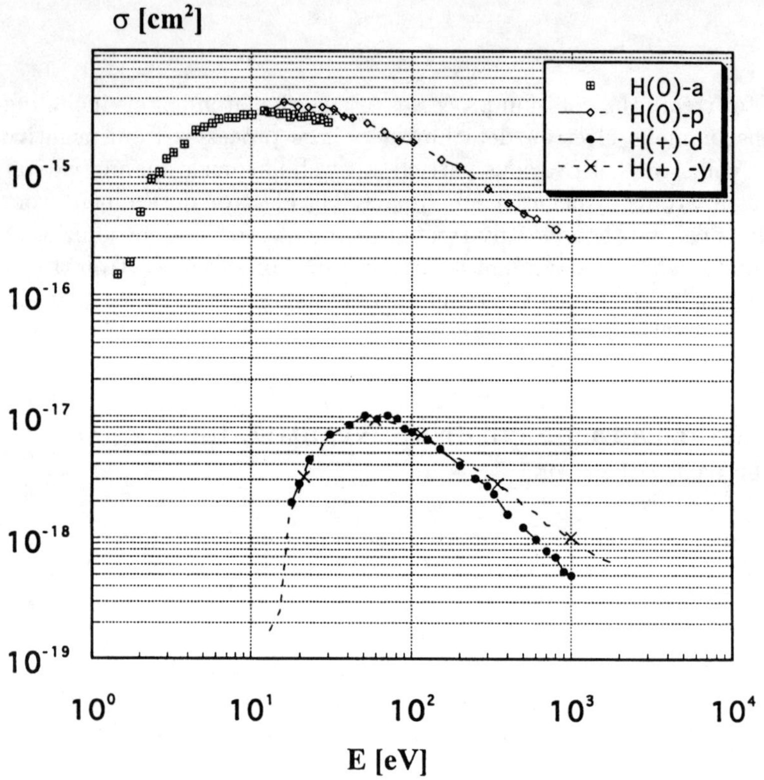

Fig. 3.1. Experimental single [3.1,5,6] and double [3.10,11] electron detachment cross sections from H^- ions under electron impact as a function of the electron energy

Though one of the previous measurements of the double-electron detachment from H^- ions, whose cross sections are roughly one order of magnitude larger than others, has been known to suffer the collisions with the trapped positive ions which are formed through the ionization of residual gases in collisions with the incident electrons and excluded here [3.9], some discrepancies still exist in the double-electron detachment cross sections even at relatively high energies [3.10, 11]. No clear reason has been found in such discrepancies as well as some structures seen at 200 eV. So far, no theoretical analysis of the double-electron detachment has been reported. An empirical formula to estimate the double-electron detachment cross sections, discussed in detail in Sect. 2.1.4, has been found to be able to reproduce quite nicely the observed data at high energies [3.12]. As seen in Fig. 3.1, the energy dependence of the impact electrons at high energies are practically the same for both the single- and double-electron detachment cross sections, suggesting strong intercorrelation between the two electrons in H^- ions.

3.1.2 Heavier Negative Heavy Ions

The single-electron detachment cross sections have been reported for typical negative ions, C^-, O^-, F^-, and found to decrease as the electron affinity of the ions (1.27, 1.47 and 3.40 eV, respectively) [3.5,13,14] increases (Fig. 3.2). Most of the cross sections observed in the single-electron detachment are in a reasonably good agreement with each other. But the observed double-electron detachment cross sections [3.15–17] are still in some disagreement among the experiments, particularly at the low-electron energy region. No experimental investigations of the multiple (more than three electron) detachment processes have been reported yet. No calculations for these cross sections have been performed, either. Only an empirical formula for the double-electron detachment has been proposed and found to reproduce the observed data within a factor of two for these heavy negative ions [3.12].

3.2 Electron Detachment from Negative Ions in Neutral Atom/Molecule Collisions

The negative ions are known to play a role in various fields such as the atmospheric physics. The importance of the multiple electron detachment processes from heavy negative ions has also been known for a long time as they are indeed required, for example, in the efficient operation of the tandem type accelerators where the injected negative ions are converted into the positive ions at the positive high voltage terminal at the megavolt (MV) through collisions with neutral gases or thin foils and accelerated back to the ground potential. These schemes have depended mostly upon experience. Yet only a limited quantitative investigations such as the cross sections have been reported so far in the MeV energy region [3.18].

A great deal of attention is being paid to the production of neutral hydrogen atoms through electron detachment processes from H^- ions due to its importance in plasma heating by injecting powerful (up to 100 MW),

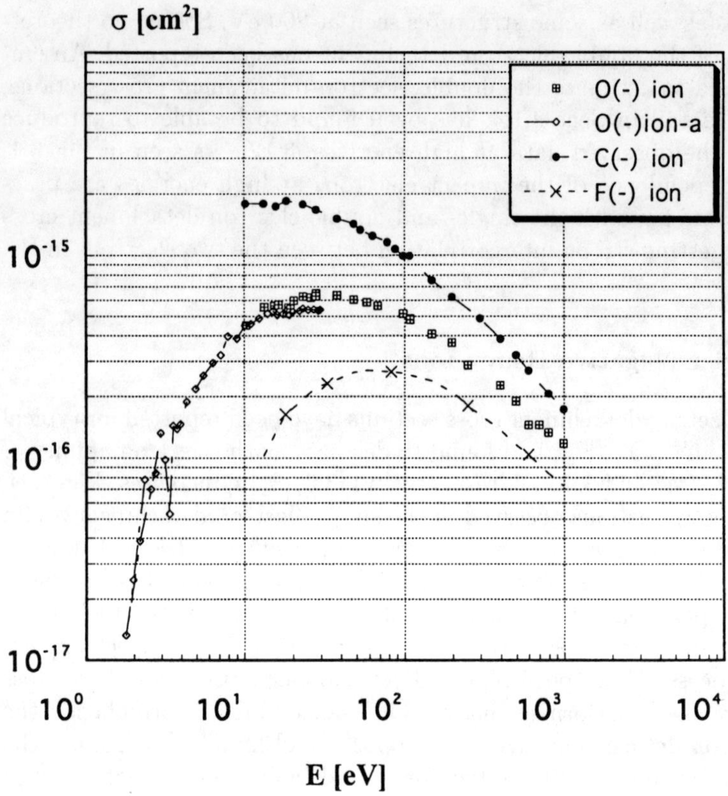

Fig. 3.2. Experimental single-electron detachment from C^-, O^- and F^- ions in electron impact as a function of the electron impact energy [3.5, 13, 14]

high-energy (100 keV–1 MeV) neutral hydrogen (H^0) atom beam into fusion plasmas, so-called Neutral Beam Injection (NBI) systems. Therefore, the conversion efficiency of H^- ions into neutral H^0 beams is one of the most crucial issues in realizing nuclear fusion reactors for energy production.

3.2.1 H^- Ions

The following single- and double-electron detachment processes:

$$H^- + B \rightarrow H^0 + e^- + \Sigma B, \tag{3.3}$$

$$H^- + B \rightarrow H^+ + 2e^- + \Sigma B \tag{3.4}$$

are the most basic collisions involving H^- ions. Here Σ indicates the summation of all the possible target states. Therefore, a lot of experimental investigations involving the electron detachment from H^- ions under collisions with neutral atoms/molecules have been reported so far. See some

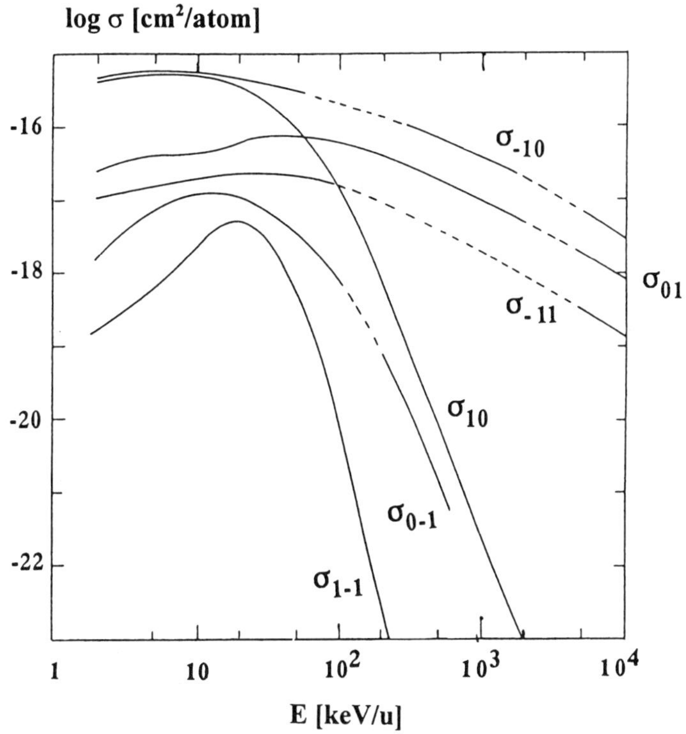

Fig. 3.3. Experimental single- and double-electron detachment cross sections of H^- ions together with the electron capture for H^+ ions and detachment (loss) from H^0 particles colliding with molecular hydrogen targets. From [3.20]

summary and review reports [3.19, 20]. Figure 3.3 exhibits a typical behavior of the cross sections as a function of the collision energy for various collision processes of hydrogen beams including the electron detachment from H^- ions under collisions with hydrogen molecules [3.19, 20]. Here σ_{ij} represents the cross sections of the electron transfer from i to j. In Fig. 3.4 are shown the summarized cross sections for the single- and double-electron detachment, σ_{-10}, σ_{-11}, from H^- ions colliding with various neutral rare gas atom targets [3.21, 22]. These figures clearly depict the collision energy dependence of the electron-detachment cross sections which has relatively slow reduction at high energies ($1/v^2$) and seems to be very similar to that of the ionization process but quite different from that of the electron transfer (capture) process which decreases quickly at higher energies ($1/v^{11}$). More new results can be found in [3.23–27].

On the other hand, the theoretical investigations on the electron detachment from negative ions are relatively limited so far, though the first

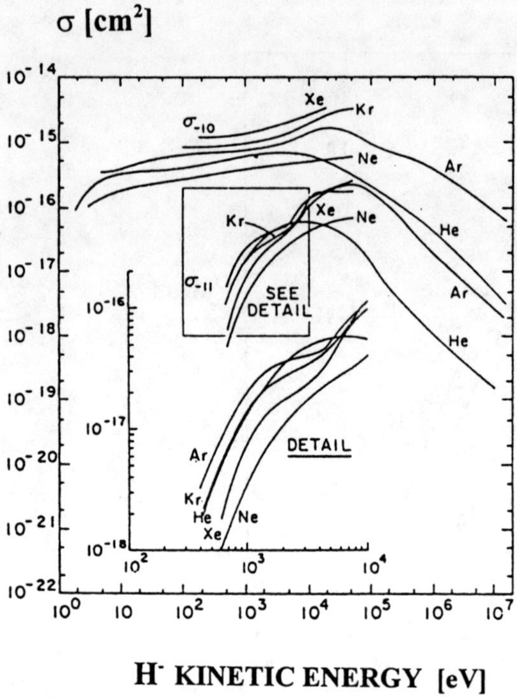

Fig. 3.4. Cross sections of the single- and double-electron detachment from H^- ions colliding with rare gas atoms. From [3.21]

theoretical treatment had been reported quite many years ago [3.28, 29]. The single-electron detachment from negative hydrogen ions colliding with neutral atoms seems to be theoretically well understood at the high energy region [3.30, 31]. Some empirical formulas for calculating the electron detachment cross sections have been introduced [3.32–34] and found to reproduce the observed data reasonably well but the double-electron detachment processes, whose observed data can not be reproduced in an independent-electron model [3.35], seem to require more elaborate treatments such as the inclusion of the electron correlation effects.

3.2.2 He^- Ions

He^- ions are known to be quite difficult to be produced with sufficient intensities and to exist only in the metastable $1s2s2p\,^4P$ states with a longest life-time of about 350 μs. Therefore, only a few experimental data are available. Figure 3.5 exhibits the cross sections for the electron detachment from He^- ions under collisions with various gas atoms over a limited collision energy region [3.36, 37]:

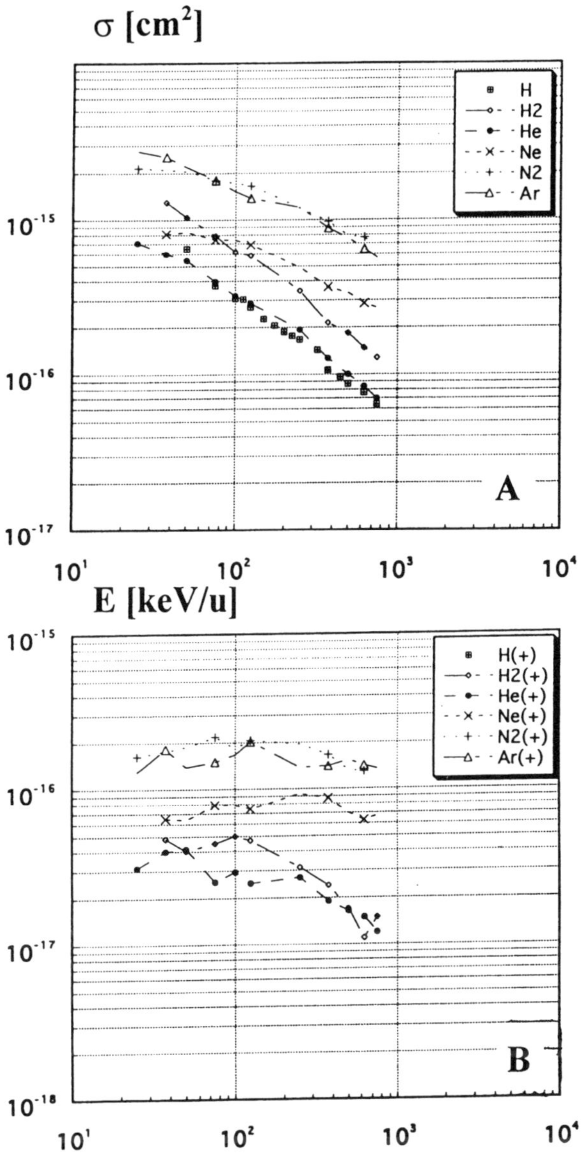

Fig. 3.5. Cross sections for the single-electron (A) and double-electron (B) detachment of He^- ions colliding with atoms and molecules [3.36]

$$He^- + B \rightarrow He^0 + e^- + \Sigma B, \qquad (3.5)$$
$$He^- + B \rightarrow He^+ + 2e^- + \Sigma B. \qquad (3.6)$$

It is seen that low-Z atomic targets (H, He) have small cross sections for both single- and double-electron detachment, meanwhile they are generally large for high-Z atoms (for example, Ar or large molecules N_2). It is interesting to note that the cross sections for the double-electron detachment decrease unexpectedly more slowly than those for the single-electron detachment. It is also remarkable that the observed ratios of the single-electron detachment from both H^- and He^- ions in atomic and molecular hydrogen targets approach 0.5 at the collision energies higher than 0.25 keV/u, indicating that a molecular hydrogen behaves like two independent hydrogen atoms. So far, there is no report on the investigations of the triple-electron detachment from He^- ions:

$$He^- + B \rightarrow He^{2+} + 3e^- + \Sigma B. \qquad (3.7)$$

It is important to measure these cross sections over a wide range of the collision energy in order to understand the detachment processes from He^- ions.

3.2.3 Heavier Negative Ions

A limited number of the cross sections involving heavier negative ions have been measured at low energies below 1 keV/u [3.38, 39]. A review on this aspect has been given [3.40]. Practically no systematic measurements of the electron detachment cross sections at higher energies have been performed. It would be important to systematically measure the electron detachment cross sections for various heavier negative ions as the most elements in the periodic table, except for the rare gases, have been found to form negative ions with sufficient intensities under some surface conditions, such as the Cs-coated surfaces.

3.2.4 Secondary Electrons from H^- Ions under Neutral Atom Collisions

The secondary electrons emitted from H^- ions under collisions can also provide an important information on the detailed mechanisms involving H^- ion collisions [3.41]. The zero-degree electron spectroscopy method (Chap. 1) becomes a powerful technique. In such collisions there are two possibilities: (i) single electron-detachment and (ii) double electron-detachment. Using the charge-selected projectile-secondary electron coincidence techniques, the secondary electrons from these two processes can be separated. In the single-electron detachment in coincidence with neutral hydrogen atoms, the electron is produced not only through the direct electron detachment

$$H^- + B \to H^0 + e^- + \Sigma B, \tag{3.8}$$

but also through the shape resonance involving the doubly excited ^1P state of H^- ions which decays via the autodetachment process:

$$H^- + B \to H^{-**}(2s2p\,^1P) + \Sigma B \to H^{0*}(n = 2) + e^- + \Sigma B. \tag{3.9}$$

The former has the cusp-shape with a peak at the electron energy corresponding to the velocity of the projectile H^- ions [3.42–44]. At the same time, the electrons emitted from the shape resonance along the forward and backward directions with respect to the projectile incident direction result in the two roughly symmetric peaks which are located on the wings of the cusp, as seen in Fig. 3.6a. Relative intensities between these two processes are found to depend on the collision partner and collision velocity. These resonance peak intensities do not depend much on the emitted angles but the cusp-peak disappears quickly at larger angles. There are distinct differences in the energy spectrum of the electrons emitted from the double-electron detachment processes

$$H^- + B \to H^+ + 2e^- + \Sigma B, \tag{3.10}$$

which shows only a single cusp-peak at the corresponding projectile ion velocity (Fig. 3.6b), but clearly asymmetric shapes with the preferable forward emission which are understood to be due to the strong correlation of two electrons under the Coulomb field of the proton. But no detailed theoretical analysis of the asymmetric shape and the correlation of two electrons in the double-electron detachment processes has been reported. Very similar behavior has been observed both at low and high energies. It should be noted that as the projectile ion velocity increases, the ionization channels of the target atoms resulting in the continuous electron emissions become significant.

3.3 Electron Detachment from Negative Ions in Positive-Ion Collisions

This is one of the recent theoretical and experimental topics in investigations of the electron detachment from negative ions as the strong Coulomb field of positive ions, in particular in highly charged ion impact, is expected to play a role. Here the electrons in H^- ions can be detached either into the projectile states (electron capture or mutual neutralization) or into the continuum. During the single-electron transition collisions of H^- ions with

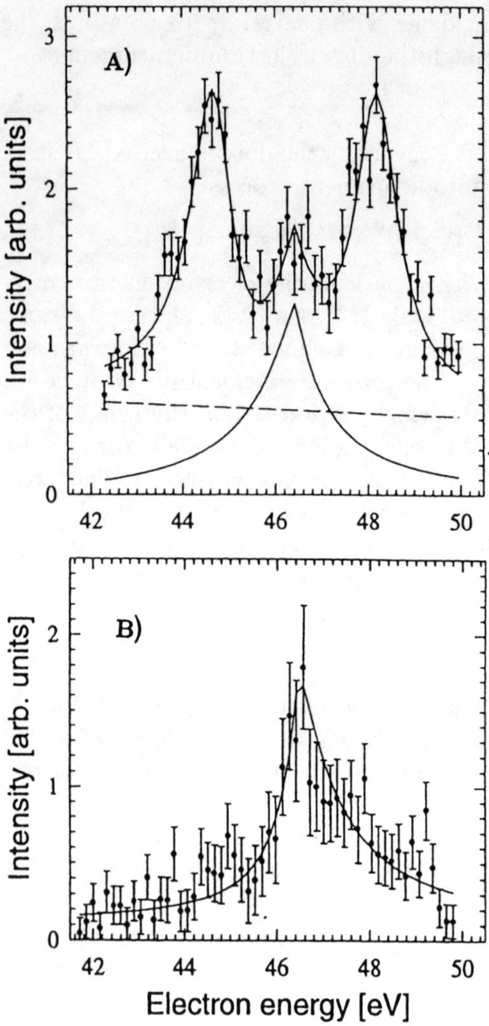

Fig. 3.6. Secondary electron energy spectra emitted from 85 keV H^- ions coincident with single-electron (a) and double-electron (b) detachment processes in collisions with neutral atom targets. From [3.45]

positive ions, the loosely bound, $1s'$- electron with the binding energy of 0.75 eV in H^- ions is active, while the inner, 1s-electron can be considered as a spectator. Thus, the active electron experiences the weak short-range potential of the screened neutral core of H(1s) and the long-range, attractive Coulomb potential of the positive ion. Generally, the electron transfer is dominant below the collision energy of 1.4 keV/u, corresponding to the orbital velocity of the the active electron. On the other hand, above that energy, the active electron is coupled to the continuum and the interactions result in the ionization.

Another and urgent issue is related with the neutral beam injection systems in fusion plasma applications where it has been known for a long time that the maximum conversion efficiencies of H^- ions to H^0 colliding with neutral atoms can never exceed 55–60 % based upon the known electron detachment cross sections (Fig. 3.3). This simply means that roughly 40–45 MW of the power of the initial negative H^- ion beams is lost during the conversion to H^0 beam if the total power of negative H^- ions is 100 MW. Some new techniques should be developed to overcome this problem. One of the ideas is the use of plasmas as a conversion medium instead of neutral atoms [3.46, 47]. The most important process here is collisions with positive ions in plasmas where the electrons play a very little role as they are free and generally relatively cold. Of course, in H^- ion collisions at high energies above a few 100 keV, the electron detachment from H^- ions can also be caused by the direct ionization in collisions with these cold electrons, as described in Sect. 3.1.

3.3.1 Proton Impact

The most basic collisions between H^- ions and positive ions is related with proton impact. The single-electron transfer (ST) process, often called mutual neutralization process

$$H^- + H^+ \to H^0(1s) + H^0(n), \tag{3.11}$$

is dominant at low energy collisions, where n represents the principal quantum number [3.48–52]. The electron is, from the energy correlation diagram of molecular hydrogen, expected to be captured into $n=$ 2, 3 and 4 excited states of hydrogen atom and is known to be the most important loss mechanism of H^- ions in astrophysical plasmas. Most of the earlier measurements had shown some structures at the collision energy below 100 eV which were assumed to be due to the formation of the quasi-molecule, H_2^* [3.51, 52]. But recent precise measurements [3.53] have indicated that there is no such structure but the cross sections are varied smoothly down to 1 eV as a function of the collision energy, as expected theoretically (Fig. 3.7).

Though the earliest theoretical treatment was based upon the Landau-Zener formula [3.54], new calculations, based upon more accurate atomic orbital [3.55, 56] and molecular orbital [3.57–59] expansion methods with the common translational factor, show a reasonable agreement with the experimental data for the mutual neutralization as well as for the single electron detachment [3.59] at the velocities of $v < 1$ a.u.$=$ 2.2 $\times 10^8$ cm/s. At low-to-intermediate energy region, the *transfer ionization* (TI) process

$$H^- + H^+ \to H^+ + e^- + H^0 \tag{3.12}$$

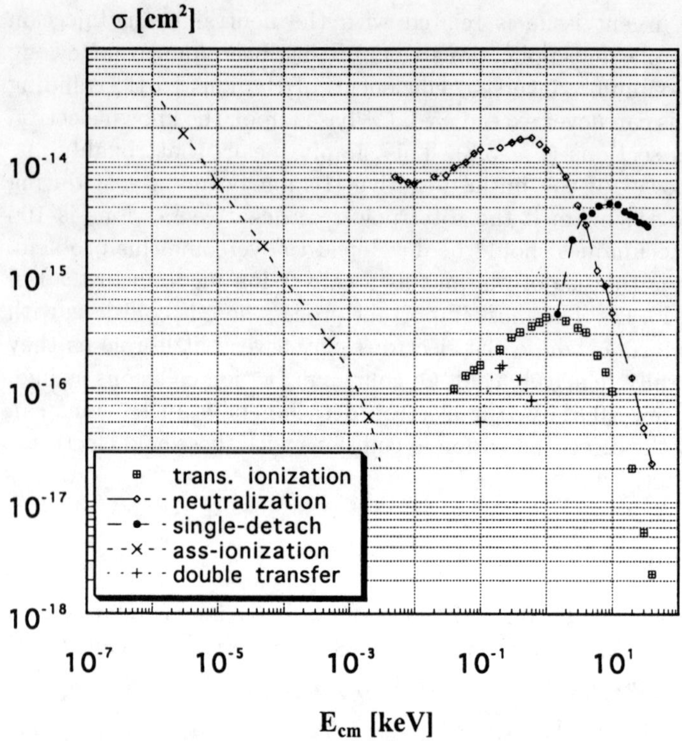

Fig. 3.7. Single-electron transfer (mutual neutralization) cross sections in $H^- + H^+$ collisions. Those for other processes are also included. The references are as follows: ST – [3.53], TI – [3.60], SD – [3.61], AI – [3.64], DT – [3.62,63]

plays some role, whose cross sections are more than one order of magnitude smaller than those for the mutual neutralization process (3.10) at the intermediate energies, as shown in Fig. 3.7 [3.60]. On the other hand, at the high collision energy region the single electron detachment (SD) process

$$H^- + H^+ \rightarrow H^0 + e^- + H^+ \tag{3.13}$$

has been confirmed experimentally to be dominant [3.61]. At 3.5 keV, the cross sections for the single-electron detachment becomes equal to those for the mutual neutralization and indeed are far dominant at higher energies (Fig. 3.7).

At higher energy proton impact, the double-electron detachment process

$$H^- + H^+ \rightarrow H^+ + 2e^- + H^+ \tag{3.14}$$

can become important but no experimental results have been reported yet.

The *double-electron transfer* (DT) process

$$H^- + H^+ \to H^+ + H^- \tag{3.15}$$

has been studied only at low energies (up to 570 eV) [3.62, 63], as shown also in Fig. 3.7. The observed results are found to be roughly two orders of magnitude smaller than those for the mutual neutralization due to the single electron transfer process (3.10) and reveal some structures, though not clear in this figure, which seem to be associated with the formation of a quasi-hydrogen molecule during collisions. It is necessary to have more sophisticated molecular theories, including the energy diagram of hydrogen molecule involving the ionic states and the covalent states at small nuclear distances in order to explain the observed structures.

The associative attachment process between H^- and H^+ ions

$$H^- + H^+ \to H_2^* \to H_2^+ + e^- \tag{3.16}$$

usually forms the excited state of a molecular hydrogen which, in turn, emits an electron and finally results in the formation of molecular ion. This process has been found to be dominant at very low energies [3.64] (Fig. 3.7).

3.3.2 Heavy, Low-Charged Positive-Ion Impact

More cross sections for the single-electron transfer involving different singly charged ions have been measured. A comparison of these cross sections for H^+, He^+ and Li^+ ions indicates that the cross sections strongly depend on the potential energy of the target positive ion at very low energies [3.65–67], indicating that the quasi-molecular formation may play a role, meanwhile at the collision energies higher than 1 keV/u they become dependent only on the target ionic charge but independent of the target ions (Fig. 3.8).

The mutual neutralization between different ions, for example,

$$H^- + He^+ \to H^0(1s) + He^{0*}(1s, n\ell m) \tag{3.17}$$

has been precisely studied at low energies using an "animated" merged beam technique [3.68] where both ion beams are two-dimensionally scanned (Chap. 2). This process has also been theoretically studied and shown [3.68, 69] that the Coulomb attraction between the negative ions and positive ions results in the sharp increase of the cross sections at low energies. Indeed the detailed molecular-orbital close-coupling calculations of the cross sections at low energies show that the most dominant channel is $n = 3$. Another important collision involving positive molecular ions such as

$$H^- + H_2^+ \to H^0 + H_2^0 \tag{3.18}$$

is interesting as both ions usually co-exist in a cold plasma. At low energies, the mutual neutralization is dominant and the cross sections at 5

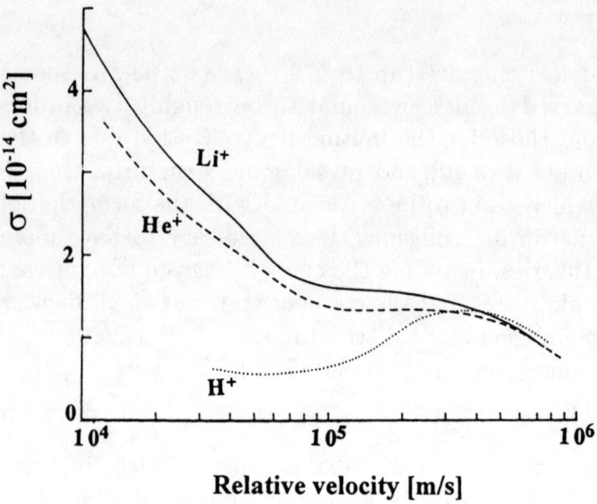

Fig. 3.8. Mutual neutralization cross sections in collisions of H^- ions with singly charged positive ions. From [3.67]

eV exceed 5×10^{-14} cm^2 [3.48]. It was found that the observed cross sections depend on the incident energy of H_2^+ ions. This may indicate that the incomplete collection of the scattered, neutralized molecules which may be formed through the dissociation of short-lived complex H_3^* molecules. It should also be noted that the internal energy (rotational, vibrational and electronic excited states) of the primary hydrogen molecular ions should strongly influence this process. In investigations with the internal-energy state specified, hydrogen molecular ions should provide more insight in such collisions involving molecular ions [3.70].

3.3.3 Highly-Charged, Heavy-Ion Impact

As the binding energy of the loosely bound electron in H^- ions is very small, the electron is captured into high Rydberg states of the positive ions in collisions of H^- ions with highly charged ions which is similar to the situations in highly charged ion and neutral atom collisions. Also, it is expected that the double-electron capture may result in the formation of the autoionization states.

Very few experimental investigations involving highly charged positive ion collisions have been performed so far, as described later in this section.

The simplest, single-electron transfer (capture) collision between H^- ion and bare, heavy ion He^{2+}:

$$H^- + He^{2+} \rightarrow H^0 + He^+(n) \tag{3.19}$$

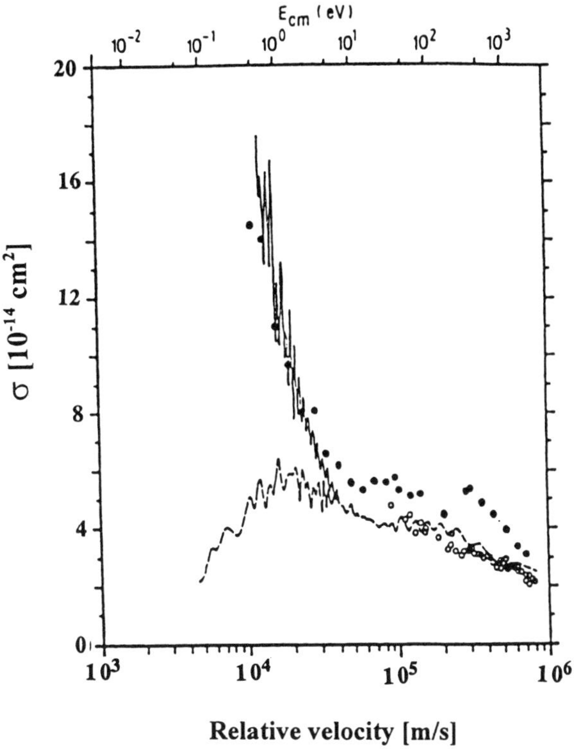

Fig. 3.9. Cross sections for the single-electron transfer (3.20) in $H^- + He^{2+}$ ion collisions. The transfer ionization (3.24) cross sections have similar energy dependence but are roughly three orders of magnitude smaller than those for the single-electron transfer shown above. The experimental data are displayed with solid and open circles. The solid and dashed curves correspond to the two-state quantum and the straight-line quasi-classical 5-state calculations, respectively. From [3.72, 73]

has been studied experimentally [3.71, 72] and theoretically [3.73] in details. From the energy correlation diagram and quasi-molecular theoretical calculations, the electron is found to be captured dominantly into the $n = 5$ state at low energies, meanwhile the capture into $n = 4$ becomes important at higher energies. The observed cross sections are shown in Fig. 3.9 where a number of the oscillations in the theoretical cross sections are understood to be due to many crossings between the incident ionic and the outgoing covalent channels.

None of the following processes have been investigated yet:

$$H^- + He^{2+} \rightarrow H^0 + e^- + He^{2+} \tag{3.20}$$

$$H^- + He^{2+} \rightarrow H^+ + 2e^- + He^{2+} \tag{3.21}$$

$$H^- + He^{2+} \rightarrow H^*(n) + He^+(n') \tag{3.22}$$
$$H^- + He^{2+} \rightarrow H^+ + He^0(n,n'), \tag{3.23}$$

i.e., single- and double-electron-detachment (3.20), (3.21), transfer excitation (3.22) as well as double-electron transfer (3.23) processes.

It should be noted that in the double-electron transfer process (3.23), both electrons are expected to be captured into the excited states, thus forming the doubly excited system which can be autoionized. The cross sections for the transfer ionization process

$$H^- + He^{2+} \rightarrow H^+ + e + He^+ \tag{3.24}$$

has been investigated and found to have the energy dependence which is very similar to the single-electron capture mentioned above (3.19), though smaller by three orders of magnitude, suggesting that the two-step processes seem to play a role as follows [3.74, 75]. In the first step the single-electron capture forms the excited $He^+(n = 3\text{-}5)$ state where the captured electron is far away from the helium nucleus. Thus, near the nucleus the following resonant electron capture:

$$He^+ + H(1s) \rightarrow He^{**} + H^+ \rightarrow He^+(n = 2) + e^- + H^+ \tag{3.25}$$

can occur, forming the doubly excited He^{**} states which, in turn, is autoionized.

The following single-electron transfer processes involving highly charged ions

$$H^- + B^{2+} \rightarrow H + B^+, \tag{3.26}$$
$$H^- + C^{3+} \rightarrow H + C^{2+} \tag{3.27}$$

have also been investigated up to 3–5 keV/u [3.76] and the results indicate that the electron capture cross sections increase as the nuclear and atomic charge of the target ions increases, as compared with those under He^{2+} ion impact (Fig. 3.10).

Investigations of H^- ion collisions with more highly charged ion impact having the charge $q \geq 4$ are still limited:

$$H^- + B^{q+} \rightarrow H + B^{(q-1)+}, \tag{3.28}$$
$$H^- + B^{q+} \rightarrow H + e^- + B^{q+}. \tag{3.29}$$

The first reliable single and double-electron detachment cross sections from H^- ions at 50 keV/u $Ar^{q+}, q = 1\text{–}8$, ions have been measured [3.47, 77]. At this energy, the single-electron detachment process (3.28) is expected to be dominant. It is found that the single-electron detachment cross sections increase relatively slowly as the projectile charge increases as $q^{1.3}$, in contrast to a simple expectation for the ionization process [3.78] and the double-electron detachment cross sections are roughly a factor of 20 smaller than

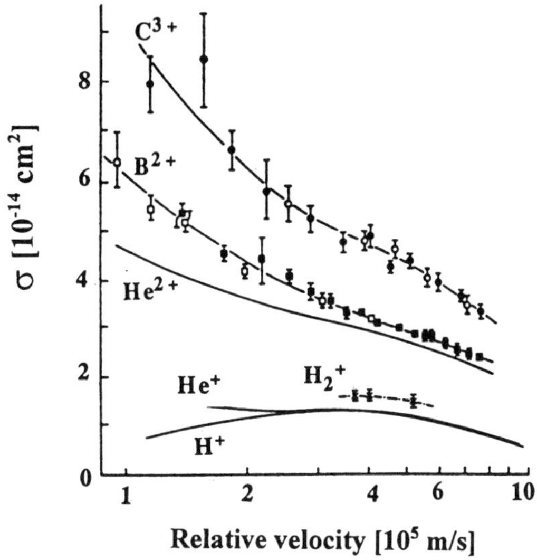

Fig. 3.10. Cross sections of the single-electron detachment from H- ions under multiply charged positive ion impact. From [3.76]

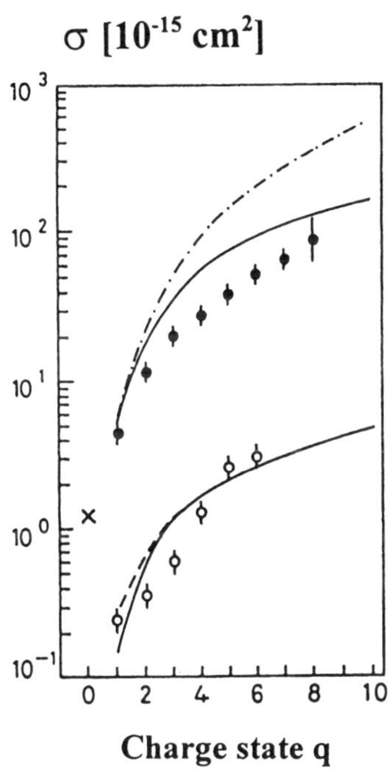

Fig. 3.11. Ar^{q+} projectile charge dependence of the single-(upper curves) and double-(lower curves) electron detachment cross sections from H^- ions at the center-of-mass kinetic energy of 50 keV. The solid and dot-dashed curves correspond to the CTMC and Bethe-Born approximation, respectively. The dashed line corresponds to the electron detachment from H^0 particles. From [3.77]

those for the single-electron detachment (Fig. 3.11). Their observation for 3–100 keV Ar^{4+} ion impact shows a very weak collision energy dependence of the electron detachment cross sections (Fig. 3.12).

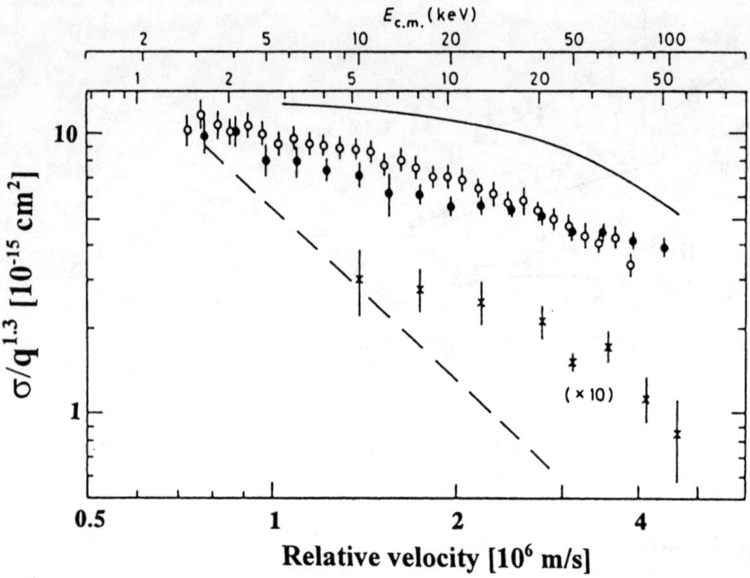

Fig. 3.12. Collision-energy dependence of the scaled single-electron detachment cross sections (solid circles) from H^- ions in $H^- + Ar^{4+}$ collisions and the scaled double-electron detachment cross sections (crosses) from H^- ions in $H^- + Ar^{3+}$ collisions (multiplied by a factor of 10). The solid and dashed curves correspond to the CTMC calculation and an E^{-1}-dependence, respectively. For comparison, those (open circles) for $H^- + H^+$ collisions are also shown. From [3.77]

These and other experimental results up to 200 keV for the single-electron detachment process can be explained reasonably well using the Keldysh theory of multiphoton ionization [3.79, 80]. The following expression has been derived over a wide range of the parameters such as the projectile ion charge and collision energy (in 10^{-16} cm^2 units):

$$\sigma_{-10} = 23.9q \ln(10q/v^2 + 0.8), \quad v^2/q \le 1.2, \tag{3.30}$$

$$\sigma_{-10} = 51.6(q/v)^2 \ln\left[\frac{2.03v^2/q}{(1+0.139v^2/q^2)^{1/2}} + 1.01\right], \quad v^2/q \ge 1.2, \tag{3.31}$$

Here the projectile velocity, v, is given in the atomic units. For $v \gg 2q$, the equations above agree with the ab initio Bethe-Born expression for the ionization by bare projectile ions, though the latter tends to overestimate the cross sections even at 200 keV.

Figure 3.13 shows a comparison between the experimental data for the single-electron detachment cross sections and some theoretical calculations in the scaled parameters, namely, E/q vs σ/q [3.79]. A different theoretical approach based upon a two-state model has been used to calculate the single-electron detachment [3.81].

Unfortunately, the electron transfer processes (3.28) into projectile ions, which are expected to be dominant at low energies, have not been experimentally investigated under multiply charged ion impact.

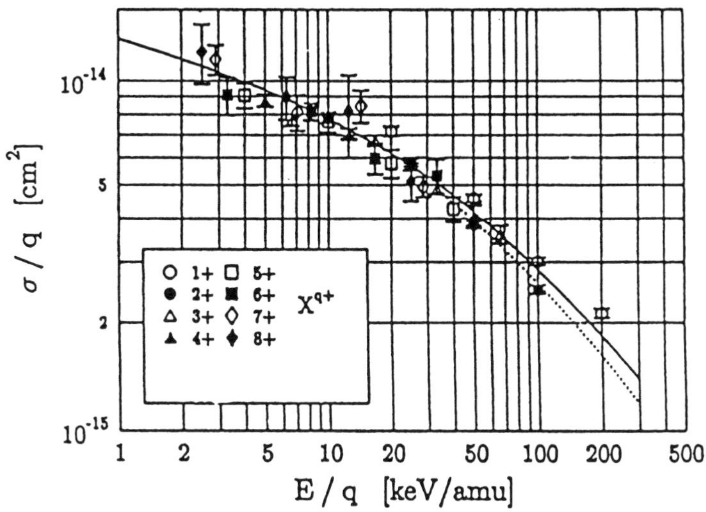

Fig. 3.13. Single-electron detachment from H^- ions under multiply charged X^{q+} ion impact as a function of the scaled energy. The solid line corresponds to (3.30) and (3.31). The dotted line is the same for $q = 1$. From [3.79]

The double-electron detachment under highly ionized ion impact has been investigated very little. The second-order perturbation theory based on the Volkov-Keldysh approach [3.82], neglecting the electron-electron correlation effect, gives the following simple expressions for the double-electron detachment cross sections from H^- ions under highly charged ion impact:

$$\sigma_{-11} = R_{so}\sigma_{-10} + \sigma_{TS}, \tag{3.32}$$

where R_{so} is the shake-off constant, being (2.3-4.0)$\times 10^{-3}$ (due to the experimental uncertainty), σ_{-10} the single-electron detachment cross section given above (3.30, 3.31) and σ_{TS} the double-electron detachment due to the two-step process [3.83] given by (in 10^{-16} cm^2 units):

$$\sigma_{TS} = 2.9(q/v)^4 \exp(-0.41q/v^2 - 1.1/v). \tag{3.33}$$

It should be noted that the main contribution in the single-electron detachment is given by the $(q/v)^2$ term, whereas that for the double-electron detachment by the $(q/v)^4$ term.

Some more systematic treatments of the single-(σ_{-10}) and double-electron (σ_{-11}) detachment from H^- ions under highly charged ions have recently been given over a wide range of the collision energy [3.80]. The scaled cross sections (σ/q) for these electron detachment processes are given as a function of the scaled collision velocity $(v/q^{1/4})$ (Fig. 3.14). These expressions can reproduce the observed data available over a wide range of collision energy. The data can be well reproduced even at low energies where the mutual neutralization plays a role [3.80].

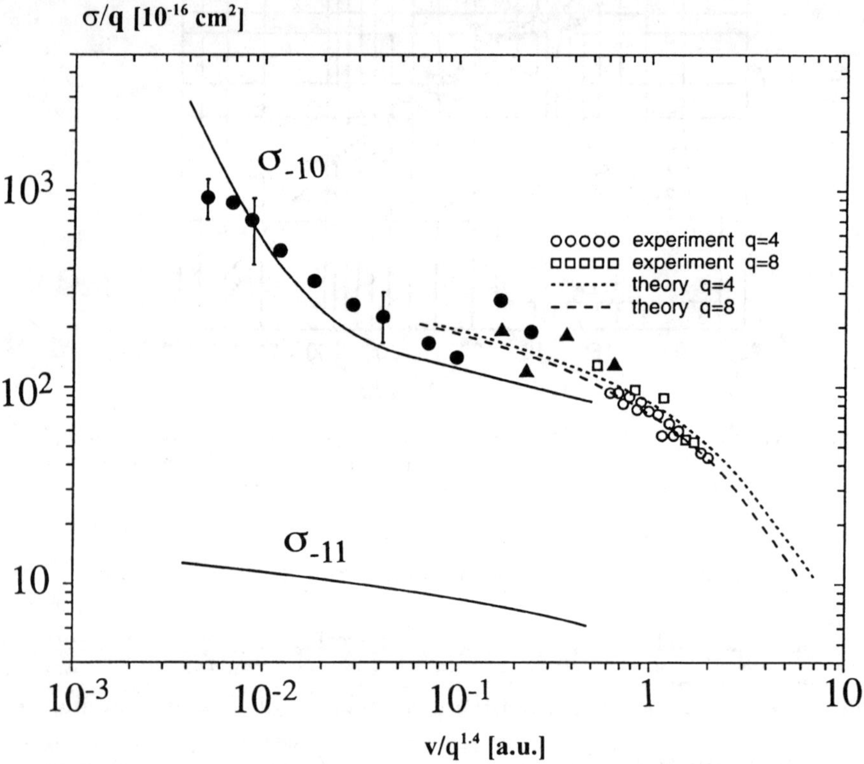

Fig. 3.14. Scaled cross sections of the single-(σ_{-10}) and double-(σ_{-11}) electron detachment from H^- ions under multiply charged ion impact. From [3.80]

3.3.4 Electron Detachment from Heavier Negative Ion

A limited information on the electron transitions in collisions of heavier negative ions, including negative molecular ions, with positive, atomic as well as molecular, ions are available only at relatively low energies [3.84–88] where the mutual neutralization is dominant. These processes at low energies are known to play a key role in the ionoshpere and in determining the balance of ion species there and also in industrial low-temperature plasmas relevant to material applications. Multiple electron-detachment cross sections from such heavier negative ions, which is of the practical importance, for example, in accelerator applications,

$$O^- + He^+, O^+,\ N^+, Na^+, N_2^{2+}, O_2^+ \rightarrow O + ... \tag{3.34}$$

have not been reported yet.

3.4 Electron Detachment in Negative Ion Collisions

In particular, the electron detachment in $H^- + H^-$ collision processes are interesting as the incident channel is repulsive, meanwhile the dominant out-going channels are neutral. The following processes can be expected to occur:

$$H^- + H^- \rightarrow H^- + H^0 + e^- \text{ (single−electron detachment)}, \tag{3.35}$$

$$H^- + H^- \rightarrow H^0 + H^0 + 2e^- \text{ (double−electron detachment)}, \tag{3.36}$$

$$H^- + H^- \rightarrow H^0 + H^+ + 3e^- \text{ (triple−electron detachment)}, \tag{3.37}$$

$$H^- + H^- \rightarrow H^+ + H^+ + 4e^- \text{ (quadruple−electron detachment)}. \tag{3.38}$$

At very low energies, the electron detachment in $H^- + H^-$ collisions can be described as the promotion of the quasi-molecular term, H_2^{2-}, in the initial channel by the Coulomb interaction into the quasi-molecular H_2^- continuum at the nuclear distance of $R = 36$ a.u., followed by the promotion into the continuum of the H_2^- quasi-molecules at $R = 18$ a.u. [3.89]. The single-electron detachment process is related with the decay of a quasi-stationary state. On the other hand, in the double-electron detachment, two processes have to be taken into account [3.83]: the direct and two-step processes. With increasing collision energy, the nonstationary treatments become necessary and approximated with the Keldysh non-stationary approach for the rearrangement collision problems [3.90]. The single-electron detachment cross sections at asymptotically high energies approach those of the ionization by electron or proton impact multiplied by a factor of three (as the projectile H^- ions can be assumed to consist of three particles, namely two electrons and a proton) (Sect. 3.5).

The cross sections for the processes above, except for the quadruple-electron detachment, have been measured over the collision energy of 2.5–100 keV [3.89, 90]. The single-electron detachment is the most dominant and larger by a factor of 5–10 than the double-electron detachment processes. Then the three-electron detachment is successively smaller than the double electron detachment by a factor of 30. The Keldysh nonstationary approach [3.90] is found to reproduce the observed results for the single- and double-electron detachment reasonably well. Also, the Classical Trajectory Monte Carlo (CTMC) calculations [3.77, 89] with proper model potentials can provide good agreement with the experimental results, too (Fig. 3.15).

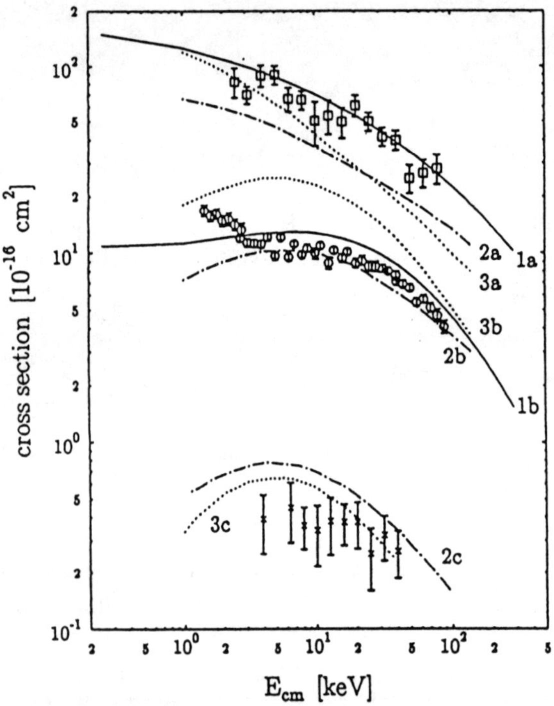

Fig. 3.15. Cross sections for the observed one-(open squares), two-(open circles) and three-(crosses) electron detachment in $H^- + H^-$ collisions. The lines with the labels a, b and c correspond to the calculated cross sections for the one-, two- and three-electron detachment processes. The labels 1, 2 and 3 represent the nonstationary tunneling approach and CTMC calculations. From [3.90]

3.5 Electron Excitation and Ionization of Neutral Atoms in Negative-Ion Collisions

Experimental and theoretical investigations of excitation and ionization by negatively charged heavy-ion impact are still scarce, except for those by electron impact. Both experimental and theoretical investigations involving multiple ionization of neutral species under negatively charged antiproton impact have been performed extensively (Sect. 3.5.3).

3.5.1 Electron Impact

As described before in Chap. 2, experimental studies on the multiple ionization under electron impact on neutral species has had a long history, starting in 1930. Yet still a number of new investigations are being reported every year. One of such continuing topics is the determination of the absolute partial ionization cross sections for different charge states as accurate as possible.

3.5.2 H^- Ion Impact

Very few investigations for excitation and ionization have been performed under H^- ion impact.

Excitation. It is expected that three particles composing of H^- ions, namely, one nucleus, H^+, and two electrons $(1s, 1s')$, may contribute to the collision processes. Indeed, at sufficiently high energies, the situation seems to be relatively simple. These three particles composing H^- ions play an independent role and, thus, the cross sections can be expressed as the sum of their contributions. It is well known that the ionization cross sections by proton impact are equal to those by electron impact at the same velocities as in the asymptotic velocity region [3.91, 92].

A limited experimental observation of the excitation [3.93] suggests that the cross sections for the excitation of rare gas atom (He) and simple molecules, (H_2, N_2) under H^- ion impact are indeed three times the corresponding proton impact cross sections over the collision energy of 0.1–2.0 MeV, though the measurements have been performed without selecting the photon energies (namely, total photon production cross sections). On the other hand, the photon emission intensities from N_2^+ first negative bands due to the excitation of N_2 molecules by 1 MeV H^- ions are found to be only slightly larger than those under proton and H^0 impact, their relative ratios being 1.27 : 1.00 : 0.61, respectively [3.94].

Ionization. The multiple ionization cross sections for rare gas atoms under H^- ion impact at 1 MeV/u region have been measured [3.95]. The observed cross sections for both the single- and double-ionization for rare gas atoms in H^- ion impact are practically the same as those by proton impact [3.96] and by electron impact [3.97, 98] with the equivalent velocities but roughly a factor of two larger than those in neutral hydrogen atoms. This indicates that at least at this energy region, the electrons in the projectile ion do not play a significant role in the ionization where the interaction between the target electrons and the projectile nucleus (e-n) is still dominant over the projectile electron-target electron (e-e) interaction [3.99]. It would be expected that the e-e interactions in the ionization processes under H^- ion impact can become important and apparent at much higher energies. But it should be also pointed out that the e-e interactions are found to play a role in the ionization of the inner-shell electrons of the heavier projectile ions where clear enhancement of the cross sections has been observed at the threshold electron velocities [3.100].

These two observations suggest that the collisions such as the excitation, where only small momentum transfer is required, occur at relatively large impact parameters, meanwhile those involving the outer-shell ionization occur at the intermediate impact parameters where the screening of the projectile nucleus by the bound electrons is important. On the other hand, those involving the inner-shell ionization occur at small impact parameters where both the bound electrons $(1s, 1s')$ may already be detached from the projectile ions, H^-, before they interact with the inner-shell electrons of target and thus the three constituting particles interact with the target electrons independently. Such a trend has already been found in experimental data of the outer-shell ionization at 1 MeV/u H^- ion impact where the ionization cross sections involving the electron detachment from H^- ions are larger than those without electron detachment from the projectile H^- ions. Still there is no consistent understanding on the role of the electrons in the projectile H^- ion impact in various collision processes whether they behave either like the screening of the projectile nucleus or like the independent particles with the same velocity of the projectile ions.

On the other hand, the situation is much more complicated at low energies because the quasi-molecular formation plays a role in every collision process [3.41]. The measured cross sections on $Na(3s \to 3p)$ excitation over 1-25 keV H^- ion impact [3.101], where the direct excitation process is dominant, are slightly smaller than those by proton and electron impact, which are nearly equal each other except for those below the threshold region, though the difference is not so large. In contrast, those in H^0 impact are much smaller, except for those below 1 keV region, than those in charged particles (H^+, H^- and electrons) where the interactions with the dipole moment of the targets play a role.

3.5.3 Antiproton Impact

A limited number of the experimental investigations have been reported on the single- and multiple-ionization of atomic and molecular targets under antiproton impact [3.102–105]. Some theoretical analysis [3.106–108] has also been performed to understand the unique features in antiproton-impact collisions resulting in multiple ionization of atoms. Recent progress in antiproton and proton impact have been reviewed in detail [3.102]. Figure 3.16 shows the single-electron (a) and double-electron (b) ionization of He by antiprotons and also the ratios (c) of the double-to-single ionization cross sections. Indeed, these ratios of He ionization at 10 keV antiproton impact have been measured to be more than 0.1 which is roughly two orders of magnitude larger than that observed at asymptotically high energies. This is in sharp contrast with that at proton impact where the ratios at 10 keV are much smaller than the asymptotic values [3.83].

The most interesting phenomenon is the fact that the ionization cross sections, in particular those for the multiple ionization, under antiproton impact at the low-intermediate energies become significantly larger than those under proton impact (due to Z_p^3 term, so-called Barkas effect [3.103]. This can be qualitatively understood from the difference in the Coulomb deflection near the target nucleus and also in the distortion/polarization of the electron orbits in the target.

Recent calculations [3.109,110] revealed the cross sections for the double ionization of He atoms at 10 keV are roughly two orders of magnitude larger in antiproton impact than in proton impact, meanwhile those above 50 keV are nearly equal, resulting in the same ratios of double-to-single electron ionization cross sections in both proton and antiproton impact.

Experimental results involving multiple ionization of other gases suggest trends similar to those in He. More significant enhancement of the inner-shell ionization by antiproton impact, compared with that in proton impact, has been predicted at low energies [3.111].

It would be interesting to compare the cross sections for the electron detachment from negative ions under antiproton impact with those under proton impact in order to see the Baskas effect more clearly. Theoretical investigations on the electron detachment from H^- ions under antiproton impact have been performed sometime ago [3.112, 113]. Unfortunately, the possibilities to do the crossed-beams experiments seem to be far remote at present [3.103, 104].

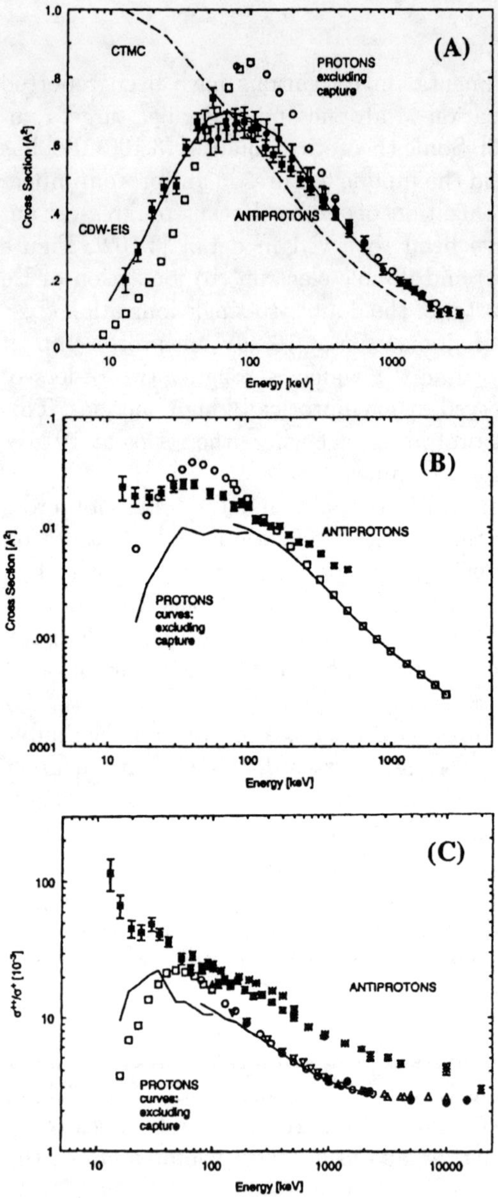

Fig. 3.16. The observed single-ionization (A) and double-ionization (B) cross sections of He under antiproton impact and ratios (C) of double-to-single ionization cross sections. In (A), the closed and open symbols represent those for antiproton impact and those for proton impact excluding the electron capture, respectively. The solid and dashed lines correspond the continuum-distorted wave-eikonal initial state calculations and CTMC calculations, respectively. In (B), the closed and open symbols represent those for antiproton impact and those for proton impact including the electron capture, respectively. The solid lines represent those for proton impact excluding the electron capture. In (C), the same as those in (B). From [3.102, 105]

4. Multiple Photoionization

This chapter treats multielectron photoionization processes arising in collisions of atoms and ions with a single high-energy photon. *Multiphoton multielectron* ionization of an atomic system, e.g., by an intense laser radiation, is also possible but this process is not considered here. Multiple photoionization is a subject of special attention because it is extremely sensitive to the the details of atomic structure and the correlation effects between atomic electrons. The main attention is paid to the double-ionization processes and, especially, to the ratio of double to single photoionization cross sections in He atom and He-like systems. Double-electron processes of photoionization with simultaneous excitation of the target electrons are also considered.

4.1 Basic Properties and Relations

Multiple photoionization (MPI) of an atom or ion can occur via the usual *photoeffect*

$$\hbar\omega + \mathrm{A} \rightarrow \mathrm{A}^{m+} + me^-, \quad m \geq 2, \tag{4.1}$$

when the incident photon is absorbed, or via *Compton scattering*

$$\hbar\omega_0 + \mathrm{A} \rightarrow \hbar\omega_1 + \mathrm{A}^{\mathrm{m}+} + me^-, \quad m \geq 2, \tag{4.2}$$

with the photon not being annihilated; here m is the number of the ejected electrons. The photoeffect dominates at relatively low photon energies when the photon momentum k is smaller than the average momentum p_{el} of the bound electrons: $k \ll p_{el}$. If $k \geq p_{el}$, photoeffect and Compton scattering cross sections begin to be of comparable size, and with a further increase in photon energy ($k \gg p_{el}$), the Compton scattering begins to play a key role (Sect. 4.2.2).

Similar to multiple ionization of atoms and ions by electrons (Sect. 2.1), the measured threshold energy $\hbar\omega_{th}$ for m-fold photoionization coincides with the minimal energy I_m required to remove m outermost electrons from the target A^{q+}:

$$\hbar\omega_{\text{th}} \approx I_m = \sum_{q'=q}^{q+m-1} I_{q',q'+1}, \tag{4.3}$$

where $I_{q',q'+1}$ is the one-electron ionization energy from the charge q' to $q'+1$. The $I_{q,q+1}$ energies can be estimated from the tables of *Lotz* [4.15] and *Carlson* et al. [4.16]. A table of the evaluated ionization energies I_m for some atoms is given in Sect. 2.1.

The *total photoabsorption* cross section is defined as

$$\sigma_{tot}(\omega) = \sum_{m=1}^{N} \sum_{t} \sigma_m^{(t)}(\omega), \tag{4.4}$$

where $\sigma_m(\omega)$ is the m-electron photoionization cross section and N is the total number of the target electrons. The sum on t runs over all subshells of the target. As a rule, the main contribution to the total cross section (4.4) comes from the single photoionization. Figure 4.1 shows the contribution of different processes in collisions of the He atom with a photon.

Data on the MPI cross sections give valuable information on the dynamic effects of electron–electron correlations. While values for single and total photoabsorption cross sections are given elsewhere [4.4–9], the data on MPI cross sections are rather scarce and known mainly from measurements of $\sigma_m(\omega)$ with $2 \leq m \leq 4$ in rare-gas atoms, He, Ne, Ar, Kr and Xe ([4.10–15] and references therein), O [4.16, 17], N [4.18] and in some atomic vapors ([4.19, 20] and references therein). As for ionic targets, to our knowledge, only double ionization cross sections for H^- [4.21], K^- [4.22] and Ba^+ [4.23] ions have been measured.

The most accurate measurements of the photoionization cross sections have been carried out using a synchrotron radiation and applying photoabsorption, photoelectron and photoion spectrometries [4.3, 24]. Many results have been obtained by measuring the yields of highly charged ions produced from photoionization and then converted into MPI photoionization cross sections using the reliable total photoabsorption data.

The theory of MPI processes has been successfully developed mainly for double photoionization processes [4.25, 26]. Several approaches have been used for calculations of $\sigma_2(\omega)$; among them are the many-body perturbation theory (MBPT) [4.20, 27–30], close-coupling method [4.31, 32], R-matrix approach [4.33, 34], distorted-wave approximation (DWA) [4.35–39] and variational method [4.40]. The photoionization cross sections $\sigma_2(\omega)$ have been calculated in the length, velocity and acceleration forms (see below), and depending on the choice of the initial and final wave functions, the three forms give, in general, rather different results [4.28, 30, 41]. The present status on theory and experiment on the triple *differential* cross sections of He and rare gases has been given in [4.42].

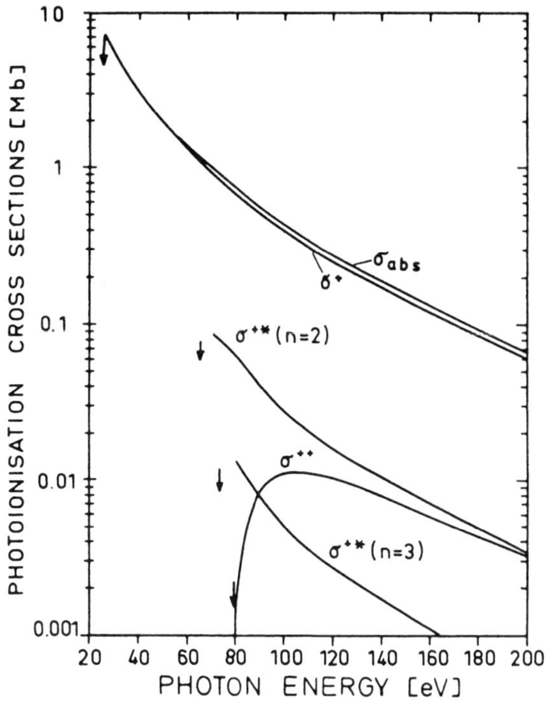

Fig. 4.1. Contribution of different processes to the total photoabsorption cross section of He (experiment): σ^+ – single ionization, σ^{++} – double ionization, $\sigma^{+*}(n)$ – ionization with simultaneous excitation to the state with the principal quantum number n. The arrows indicate the ionization thresholds. From [4.3]

When an atom absorbs an x-ray photon, multiple photoionization can occur via the following processes:
(i) two or more outermost electrons can be simultaneously ejected by a single-photon collision, so-called *direct* photoionization. This process is a one-step process important at low photon energies. Also, an inner-shell electron can be ionized or excited into an autoionizing state, and the inner-shell vacancy produced is subsequently filled through Auger, double-Auger, or Coster–Kronig transitions. One-step process occur due to electron correlation effects such as core relaxation, ground-state correlation and inelastic scattering of the photoelectron in collisions with other electrons while leaving the atom;
(ii) an inner-shell electron and one or more electrons in the outer shells can be simultaneously ejected in the initial photoabsorption step.

Similar processes occur in multiple ionization of atoms by electron impact (Sect. 2.1) but there is a big difference between photo- and charge-particle ionization: a photon transfers all its energy to the target while a charge-particle transfers only a small part of it.

Photoionization cross sections and branching ratios for multiple photoionization provide valuable information on the electron correlations and display quite complicated structures related to these effects. A typical example of MPI cross sections is shown in Fig. 4.2 for Ar atoms. Below the $2p$ ionization threshold, $Ar^{m+}(m \geq 2)$ ions are produced mainly via photoionization process with simultaneous ejection of two or more electrons. At photon frequencies $\hbar\omega \geq I_{2p,2s}$, i.e. higher than the ionization energies of the L shell, the L-electron and one or more electrons from outer $3p$-shell are ejected to produce cross section maxima of Ar^{4+}, Ar^{5+} and Ar^{6+} ions at around 390, 700, and 1000 eV, respectively. For heavier atoms the situation is more complicated because a lot of the inner-shell electrons come into play.

Both photons and charged particles interact with the matter via electromagnetic fields. The corresponding processes, involving real and virtual photons, are strongly coupled to each other because electromagnetic field of a charged particle, moving with a constant velocity, is similar to a field of a light wave. For example, under certain conditions, the matrix elements of the bremsstrahlung and Compton scattering or multielectron ionization by a single photon and a charged particle have similar structures.

In collisions of a fast charged particle with a neutral atom, the interaction of the atom with the projectile can be substituted by its interaction with a spectrum of equivalent virtual photons caused by a charged particle. In the general case (including relativistic collisions), this electromagnetic field comprises both transverse and longitudinal photons. This procedure is known as the Weizsäcker–Williams method of equivalent photons [4.44–46].

Using the method of equivalent photons, one can obtain a relation between the cross sections for ionization of m target electrons by a single photon and a charged particle. In the dipole approximation and impact parameter ρ representation, the m-electron ionization cross section $\sigma_m(v)$ can be presented as a sum of the contribution from the distant (*dis*) and close (*cl*) collisions [4.47]:

$$\sigma_m(v) = \sigma_m^{\mathrm{dis}}(v) + \sigma_m^{\mathrm{cl}}(v), \quad \sigma_m^{\mathrm{dis}}(v) = 2\pi \int_{\rho_{\min}}^{\infty} P_m^{\mathrm{dis}}(\rho)\rho d\rho, \tag{4.5}$$

$$P_m^{\mathrm{dis}}(\rho) = \frac{c_o}{\pi^2}\frac{q^2}{(\pi\rho)^2}\int_{I_m}^{\infty} S\left(\frac{\omega v}{\rho}\right)\sigma_m(\omega)\frac{d\omega}{\omega}, \quad \rho \gg \frac{q}{v}, \tag{4.6}$$

$$S(x) = x^2\left[\mathrm{K}_0^2(x) + \mathrm{K}_1^2(x)\right], \tag{4.7}$$

where c_o is the speed of light in vacuum. Here $P_m^{\mathrm{dis}}(\rho)$ is the probability for ionization of m electrons by a q-charged particle at relative (nonrelativistic) velocity v, $\sigma_m(\omega)$ is the m-electron photoionization cross section

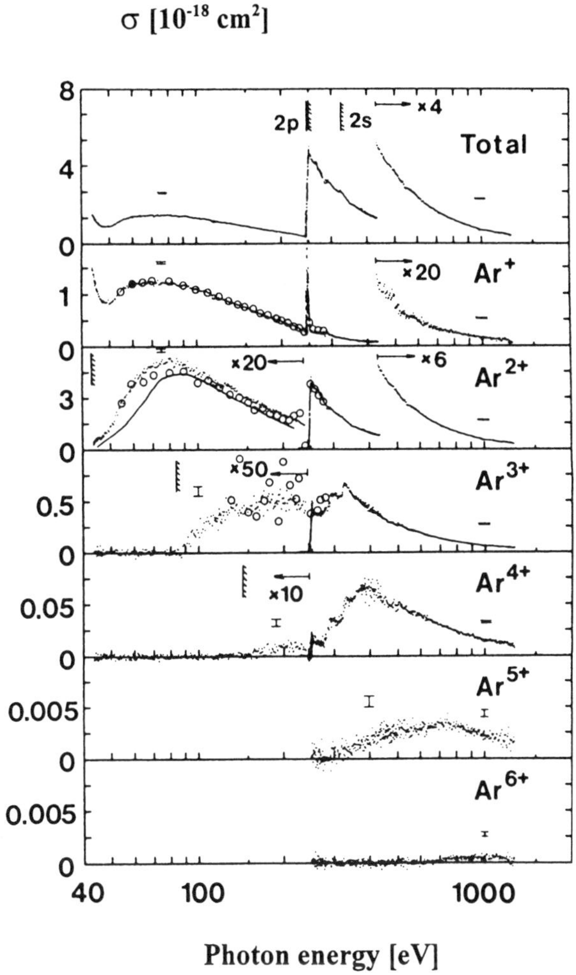

Fig. 4.2. Partial $\sigma_m(\omega)$ and total photoionization cross sections for production of Ar^{m+} ions. Experiment: open circles – [4.11], points – [4.13]; theory: solid curve for $\sigma_2(\omega)$ – [4.43]. From [4.13]

of the target at the frequency ω, $S(x)$ is a dimensionless function describing Fourier components of the electric field induced by the projectile, I_m is the ionization energy defined by (4.3), and $\mathrm{K}_a(x)$ is the modified Bessel (MacDonald) function. The minimum impact parameter $\rho_{\min}$ denotes the dividing distance which separates distant from close collisions. Integration in (4.6) over the impact parameter ρ can be easily performed analytically including the case of relativistic velocities [4.44].

Table 4.1. Recommended experimental total and double photoionization cross sections for He as a function of photon energy; $I_1 = 24.6$ eV, $I_2 = 79$ eV. From [4.14]

$\hbar\omega$, eV	$\sigma_{tot}(10^{-18}\ \mathrm{cm}^2)$	$\sigma_2(10^{-21}\ \mathrm{cm}^2)$	$\hbar\omega$, eV	$\sigma_{tot}(10^{-18}\ \mathrm{cm}^2)$	$\sigma_2(10^{-21}\ \mathrm{cm}^2)$
24.6	7.42(7)		115	0.276(4)	8.9(4)
26	7.80(7)		120	0.248(4)	8.5(4)
28	6.05(6)		125	0.228(5)	8.2(6)
30	5.38(5)		130	0.207(5)	7.8(5)
32	4.82(5)		135	0.187(54)	7.3(5)
34	4.32(4)		140	0.171(4)	6.8(5)
36	3.88(4)		145	0.155(4)	6.4(4)
38	3.50(3)		150	0.137(3)	5.7(4)
40	3.16(3)		155	0.124(2)	5.3(4)
42	2.85(3)		160	0.115(2)	4.9(4)
44	2.60(3)		165	0.104(2)	4.5(4)
46	2.38(2)		170	0.094(2)	4.1(3)
48	2.19(2)		175	0.087(2)	3.9(3)
50	2.01(3)		180	0.080(2)	3.6(3)
52	1.84(3)		185	0.073(2)	3.3(3)
54	1.71(2)		190	0.067(2)	3.0(2)
56	1.62(2)		195	0.062(1)	2.8(2)
58	1.56(2)		200	0.058(1)	2.6(2)
64	1.20(2)		205	0.053(1)	2.4(2)
66	1.15(1)		210	0.050(1)	2.3(2)
68	1.06(2)		215	0.046(1)	2.1(2)
70	0.98(1)		220	0.043(1)	1.9(2)
72	0.91(1)		225	0.0397(9)	1.8(2)
74	0.85(1)		230	0.0369(8)	1.7(2)
76	0.80(1)		235	0.0340(8)	1.5(1)
78	0.75(1)		240	0.0329(8)	1.5(1)
80	0.70(1)	1.11(3)	245	0.0309(8)	1.4(1)
82	0.66(1)	2.81(7)	250	0.0287(7)	1.3(1)
84	0.622(9)	5.2(4)	255	0.0267(7)	1.2(1)
86	0.586(9)	6.7(5)	260	0.0253(7)	1.1(1)
88	0.555(8)	7.5(5)	265	0.0238(6)	1.1(1)
90	0.523(8)	8.1(5)	270	0.0277(6)	1.02(9)
92	0.494(7)	8.6(5)	275	0.0212(5)	0.95(9)
94	0.469(7)	9.0(5)	280	0.0196(5)	0.88(8)
96	0.442(7)	9.2(5)	285	0.0184(6)	0.82(8)
98	0.421(6)	9.5(5)	290	0.0176(5)	0.78(8)
100	0.396(6)	9.8(5)	295	0.0170(5)	0.75(7)
105	0.347(5)	9.4(5)	300	0.0162(5)	0.72(7)
110	0.308(5)	9.2(4)			

Finally, the relation between the distant part $\sigma_m^{\mathrm{dis}}(v)$ and photoionization cross section $\sigma_m(\omega)$ can be written in a closed analytical form:

$$\sigma_m^{\mathrm{dis}}(v) = \int_{I_m}^{\infty} Q(\omega, v)\sigma_m(\omega)d\hbar\omega, \tag{4.8}$$

where $Q(\omega, v)$ reproduces the intensity of the frequency spectrum integrated over all possible impact parameters ρ. For low frequencies one has

$$Q(\omega, v) \approx q^2 \frac{2}{\pi}\alpha \left(\frac{c_0}{v}\right)^2 \frac{1}{\hbar\omega} \ln\left(\frac{1.123v}{\omega\rho_{\min}}\right), \quad \omega \ll v/\rho_{\min}\,. \tag{4.9}$$

Correspondingly, at high frequencies, the spectrum falls off exponentially

Table 4.2. Recommended experimental total and double photoionization cross sections for Ne as a function of photon energy; $I_2 = 63$ eV. 1 Mb = 10^{-18} cm^2. From [4.49]

$\hbar\omega$ [eV]	σ_{tot} [Mb]	σ_2 [Mb]	E [eV]	σ_{tot} [Mb]	σ_2 [Mb]
63	6.53	0.0070	345	0.276	0.0338
65	6.36	0.0269	360	0.248	0.0300
70	5.96	0.0619	375	0.220	0.0267
75	5.58	0.0982	390	0.192	0.0237
90	4.50	0.170	405	0.175	0.0218
105	3.57	0.208	420	0.160	0.0205
120	2.84	0.211	435	0.146	0.0192
135	2.24	0.201	450	0.131	0.0174
150	1.81	0.181	465	0.121	0.0156
165	1.51	0.163	480	0.112	0.0140
180	1.24	0.139	495	0.103	0.0129
195	1.04	0.121	510	0.0938	0.0118
210	0.884	0.106	525	0.0860	0.0110
225	0.780	0.0949	540	0.0802	0.0103
240	0.686	0.0845	555	0.0745	0.0097
255	0.593	0.0742	570	0.0690	0.0091
270	0.500	0.0634	585	0.0649	0.0085
285	0.429	0.0541	600	0.0611	0.0079
300	0.378	0.0481	615	0.0573	0.0074
315	0.332	0.0427	630	0.0535	0.0069
330	0.304	0.0380	645	0.0501	0.0063

$$Q(\omega, v) \approx q^2 \alpha \left(\frac{c_0}{v}\right)^2 \frac{1}{\hbar\omega} \exp\left[-\left(\frac{2\omega\rho_{\text{min}}}{v}\right)\right], \quad \omega \gg v/\rho_{\text{min}}, \tag{4.10}$$

where α is the fine-structure constant. In the ionization problem, the minimum impact parameter ρ_{min} is usually of the order of the atomic radius of the target.

Thus, according to the Weizsäcker-Williams method, a behavior of the m-electron photoionization cross sections near threshold defines the contribution from the distant collisions to the multiple ionization cross sections by charged particles. In this context, it should be noted that relativistic, highly charged ions such as 10 GeV/u U^{92+} can produce very strong electromagnetic field as high as 10^{18} W/cm^2 within a very short period of the order of 10^{-18} seconds. The photo process due to such photons could be very interesting [4.48].

The absolute experimental partial cross sections $\sigma_m(\omega)$, $1 \le m \le 4$, for rare-gas atoms are given in Tables 4.1–5.

Table 4.3. Experimental multiple photoionization cross sections for Ar [4.11]

Photon energy, eV	σ_3 [Mb] ($I_3 = 84$ eV)	σ_2 [Mb] ($I_2 = 43$ eV)	σ_1 [Mb] ($I_1 = 15.8$ eV)
55		0.136(12)	1.06(9)
60		0.194(17)	1.20(10)
65		0.201(18)	1.25(11)
70		0.220(20)	1.26(11)
80		0.224(23)	1.24(11)
90		0.227(23)	1.17(11)
100		0.198(20)	1.12(10)
110		0.200(20)	1.02(9)
120		0.184(18)	0.96(9)
130	0.010(8)	0.161(15)	0.88(8)
140	0.017(11)	0.155(15)	0.80(7)
150	0.008(9)	0.134(13)	0.75(7)
160	0.007(10)	0.119(11)	0.69(6)
170	0.010(9)	0.119(11)	0.63(6)
180	0.014(12)	0.103(10)	0.59(5)
190	0.006(11)	0.098(10)	0.53(5)
200	0.018(12)	0.086(9)	0.48(4)
210	0.013(10)	0.087(12)	0.43(5)
220	0.010(11)	0.098(13)	0.37(5)
230	0.015(11)	0.104(13)	0.32(4)
240	0.015(12)	0.147(19)	0.25(3)
250	0.378(54)	3.77(47)	0.41(6)
260	0.398(57)	3.45(4)	0.30(4)
270	0.506(69)	3.04(38)	0.29(4)
280	0.529(72)	2.69(34)	0.28(4)

4.2 Double Photoionization

Double ionization of atoms and ions by photon impact is of a current experimental and theoretical interest because it is the simplest multielectron photoionization process which allows the investigation of electron–electron correlation effects in detail.

Experimental data on the double-photoionization cross sections $\sigma_2(\omega)$ are rather scarce and, as was mentioned, are known for a few atomic systems: He, O, N, Ne, Na, Ar, Kr, Xe, H^-, K^- and Ba^+. Experimental data on $\sigma_2(\omega)$ for rare-gas atoms are given in the previous section. Since double ionization of He constitutes a special interest because helium is the simplest many-electron system, this case is considered separately in Sect. 4.2.1. Some experimental double photoionization cross sections of many-electron systems are given in Figs. 4.3, 4.

Double photoionization cross sections of Na atom and Ba^+ ion display quite a complicated structure (Figs. 4.3, 4). Figure 4.3 shows experimental

Table 4.4. Experimental multiple photoionization cross sections for Kr [4.11]

Photon energy, eV	σ_4 [Mb] ($I_4 = 128$ eV)	σ_3 [Mb] ($I_3 = 75$ eV)	σ_2 [Mb] ($I_2 = 38$ eV)	σ_1 [Mb] ($I_1 = 14$ eV)
80		0.05(1)	0.21(2)	0.384(51)
90		0.07(1)	0.39(4)	0.131(28)
100		0.27(3)	0.88(8)	0.255(42)
110		0.49(5)	1.10(10)	0.302(48)
120		0.75(7)	1.67(15)	0.218(40)
130		1.08(10)	2.1(19)	0.202(38)
140		1.38(12)	2.61(23)	0.168(36)
150		1.58(14)	2.91(25)	0.193(43)
160	0.018(2)	1.72(15)	3.19(28)	0.099(35)
170	0.025(3)	1.88(16)	3.24(28)	0.097(40)
180	0.051(6)	1.90(17)	3.24(28)	0.132(54)
190	0.074(9)	1.87(16)	3.20(28)	0.168(73)
200	0.091(13)	1.84(16)	3.06(27)	0.248(110)
210	0.154(21)	1.71(17)	2.61(24)	0.567(218)
220	0.214(30)	1.75(16)	2.65(24)	0.350(190)
230	0.233(30)	1.95(18)	2.55(23)	0.078(154)
240	0.202(26)	1.68(16)	2.37(22)	0.367(131)
250	0.206(25)	1.75(16)	2.30(21)	0.167(99)
260	0.253(30)	1.57(14)	2.24(20)	0.191(98)
270	0.216(24)	1.63(15)	2.14(19)	0.095(104)
280	0.221(25)	1.44(14)	1.88(17)	0.357(154)

value of $\sigma_2(\omega)$ for Na ($I_2 = 52.4$ eV) [4.19] measured with the 10% accuracy using synchrotron radiation and combining the results of photoelectron and photoion spectroscopies. Between 52.4 and 71.0 eV, only direct double photoionization is possible

$$\hbar\omega + \mathrm{Na}(2s^2 2p^6 3s\ ^2S) \to \mathrm{Na}^{2+}(2s^2 2p^5\ ^2P) + 2\mathrm{e}^- . \tag{4.11}$$

The absolute values of the direct photoionization cross section for the process (4.11) are given in [4.20] and are found in a good agreement with MBPT calculations.

The Fano-type peak at ~66.6 eV is related to photoexcitation of $2s$-electron into $3p$ state followed by Auger decay:

$$\begin{aligned}\hbar\omega + \mathrm{Na}(2s^2 2p^6 3s\ ^2S_{1/2}) &\to \mathrm{Na}^*(2s2p^6 3s3p\ ^2P_{1/2,3/2}) \\ &\to \mathrm{Na}^{2+}(2s^2 2p^5\ ^2P_{1/2,3/2}) + 2\mathrm{e}^- .\end{aligned} \tag{4.12}$$

At the frequency of 71.0 eV, photoionization of the inner $2s$-electron starts to lead finally to a double ionization

$$\begin{aligned}\hbar\omega + \mathrm{Na}(2s^2 2p^6 3s\ ^2S) &\to \mathrm{Na}^{+*}(2s2p^6 3s) + \mathrm{e}^- \\ &\to \mathrm{Na}^{2+}(2s^2 2p^5) + 2\mathrm{e}^- .\end{aligned} \tag{4.13}$$

Table 4.5. Experimental multiple photoionization cross sections for Xe [4.11]

Photon energy, eV	σ_4 [Mb] ($I_4 = 109$ eV)	σ_3 [Mb] ($I_3 = 64$ eV)	σ_2 [Mb] (I_2) = 35 eV	σ_1 [Mb] ($I_1 = 12$ eV)
65			1.31(12)	1.280(115)
70		0.43(14)	2.90(26)	0.948(84)
75		1.59(15)	4.76(42)	0.843(79)
80		2.51(22)	8.60(75)	0.868(82)
90		5.81(52)	16.27(1.42)	1.509(143)
100		7.49(67)	18.40(1.60)	1.111(112)
110		6.23(56)	15.84(1.39)	0.778(87)
120	0.021(3)	4.26(37)	10.01(87)	0.314(40)
130	0.025(3)	2.37(21)	5.24(46)	0.010(22)
140	0.135(17)	1.42(12)	2.49(22)	0.007(21)
150	0.306(38)	0.83(7)	1.09(10)	0.051(26)
160	0.322(40)	0.50(4)	0.51(4)	0.048(26)
170	0.294(36)	0.34(3)	0.35(3)	0.017(22)
180	0.271(34)	0.25(2)	0.36(3)	0.008(21)
190	0.196(23)	0.26(2)	0.41(4)	0.043(26)
200	0.110(19)	0.28(3)	0.57(5)	0.036(29)
210	0.152(26)	0.31(3)	0.62(5)	0.009(22)
220	0.154(26)	0.34(3)	0.74(7)	0.049(32)
230	0.089(25)	0.35(3)	0.83(7)	0.010(33)
240	0.245(42)	0.34(3)	0.69(6)	0.063(27)
250	0.202(35)	0.41(4)	0.77(7)	0.008(34)
260	0.169(29)	0.39(4)	0.83(7)	0.050(36)
270	0.235(41)	0.39(4)	0.78(7)	0.061(25)
280	0.203(36)	0.45(4)	0.81(7)	0.017(34)

Double-photoionization cross section of Ba^+ ion has been measured with the energy resolution $E/\Delta E \approx 130$ (Fig. 4.4). It displays a broad giant resonance peak below 109 eV energy corresponding to the binding energy of the inner $4d$-electron. The main photoionization process is due to photoionization of $4d$-electron followed by Auger decay:

$$\begin{aligned} \hbar\omega + \mathrm{Ba}^+(4d^{10}5s^25p^66s) &\rightarrow \mathrm{Ba}^{2+*}(4d^95s^25p^66s) + \mathrm{e}^- \\ &\rightarrow \mathrm{Ba}^{3+}(4d^{10}5s^25p^5) + 2\mathrm{e}^- . \end{aligned} \tag{4.14}$$

4.2.1 Double Photoionization of He and He-like Systems

Double photoionization of He atoms has been studied in more details as compared to other targets [4.26, 46, 50–52]. It provides a fundamental test for the correlation effects because the He atom is the simplest system to investigate the complete breakup of the bound, three-particle system into three free particles: two electrons and one nucleus. Double photoionization of He has been experimentally investigated both at low ($\hbar\omega$ = 79–560 eV)

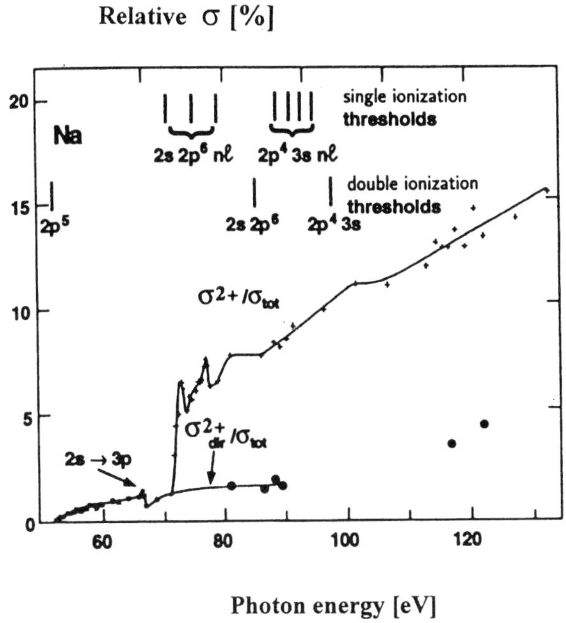

Fig. 4.3. Relative double photoionization cross section of Na measured by *Rouvellou et al* [4.19]: + double-ionization cross section (above 71 eV), × the direct double photoionization (below 71 eV); solid circles are the direct double-photoionization cross section above 71 and 110 eV, respectively. All data are normalized to the total photoionization cross section. The thresholds for single- and double-photoionizaton cross sections, $\sigma_1(\omega)$ and $\sigma_2(\omega)$, are indicated by the vertical bars

[4.10, 11, 27, 53–55] and high ($\hbar\omega > 2$ keV) [4.49, 56, 57] photon energies. Only a few measurements cover the intermediate energy range $\hbar\omega = 200$–1500 eV [4.49,58]. In case of He, most of experiments and theories are related to the ratio of double-to single-photoionization cross sections

$$R_2(\omega) = \frac{\sigma_2(\omega)}{\sigma_1(\omega)}, \tag{4.15}$$

other than the absolute double-ionization cross section $\sigma_2(\omega)$ itself.

In the nonrelativistic dipole approximation, the DPI cross section for two-electron system has the form [4.59]:

$$\begin{aligned} \sigma_2(\omega) = & 4\pi^2 \alpha a_0^2 \, \omega \; \delta(\hbar\omega - I_2 - \epsilon - \epsilon') \\ & \times \sum_l \int d\epsilon \int d\epsilon' \mid \langle \Psi_f(\mathbf{r_1}, \mathbf{r_2}) \mid \mathbf{e} \cdot \mathbf{D} \mid \Psi_i(\mathbf{r_1}, \mathbf{r_2}) \rangle \mid^2, \end{aligned} \tag{4.16}$$

where α is the fine-structure constant, a_0 is the Bohr radius, ω is the photon frequency, $\mathbf{e}$ is the direction of linearly polarized light, $\mathbf{D}$ is the dipole opera-

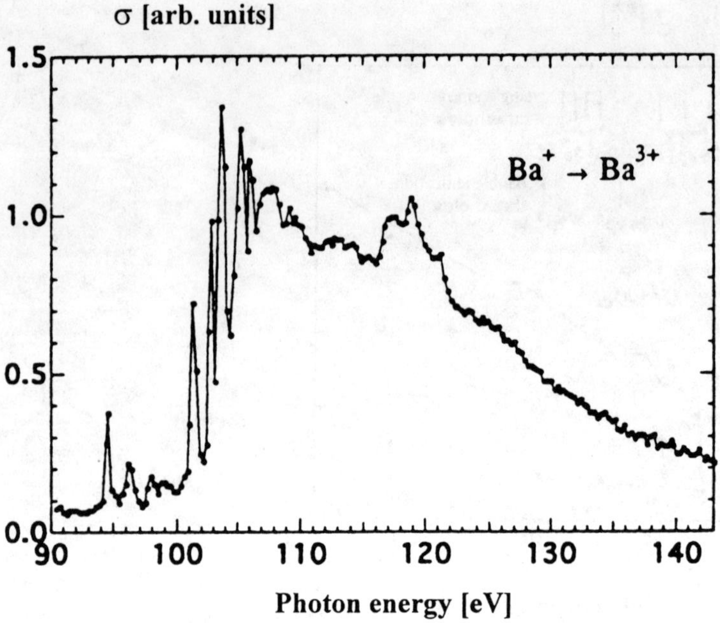

Fig. 4.4. Experimental double-photoionization cross section of Ba^+ ions [4.23]

tor and I_2 is the threshold energy for double ionization. The wave functions $\Psi_{\mathrm{i,f}}$ describe the initial ground state (i) and the final state (f) containing the nucleus and two unbound electrons with energies ϵ, ϵ', and angular momenta ℓ, ℓ', respectively. All energies are in atomic units. The momenta ℓ, ℓ' have to couple to one since the photon carries the total angular momentum $\hbar$. At high photon energies, only $\ell = 0$ and $\ell' = 1$ contribute to the cross sections [4.60]; at lower energies, higher partial waves have to be included.

For two electrons with vectors $\mathbf{r}_1$, $\mathbf{r}_2$ and impulses $\mathbf{p}_1$, $\mathbf{p}_2$, the dipole moment operator $\mathbf{D}$ can be presented in the length, velocity and acceleration forms, respectively,

$$\mathbf{D}_{\mathbf{L}} = \mathbf{r}_1 + \mathbf{r}_2, \quad \mathbf{D}_{\mathbf{V}} = \mathrm{i}\omega^{-1}(\mathbf{p}_1 + \mathbf{p}_2),$$

$$\mathbf{D}_{\mathbf{A}} = Z\omega^{-2}\left(\frac{\mathbf{r}_1}{r_1^3} + \frac{\mathbf{r}_2}{r_2^3}\right), \tag{4.17}$$

where Z is the nuclear charge of the target. These three forms are identical if the exact atomic wave functions are used.

Table 4.6. Expansion coefficients $A_{ab}^{(l)}$ for the ground-state wave function of the He atom in (4.18), (4.19). From [4.59]

(a, b)	$A_{ab}^{(0)}$	$A_{ab}^{(1)}$	$A_{ab}^{(2)}$
(0,0)	8.69519	-14.2888	-14.1186
(0,1)	2.84923	37.5271	36.4897
(0,2)	8.24437	-14.0944	-16.0409
(0,3)	-4.74978	0.264419	2.93842
(0,4)	1.36195	0.593035	-0.211587
(0,5)	-0.111747	-0.064705	0.0036127636
(1,1)	-7.73456	-35.2189	-23.3933
(1,2)	5.82892	35.2792	19.3197
(1,3)	2.35973	-5.25726	-3.197619
(1,4)	-1.27994	-0.0307959	0.197619
(1,5)	0.116503	0.0317287	-0.00270782
(2,2)	-5.00189	-9.50410	-3.45314
(2,3)	1.99741	3.00342	0.0880129
(2,4)	0.00885715	-0.110612	-0.0342703
(3,3)	-0.222887	-0.159411	-0.0314348

The cross sections $\sigma_2(\omega)$ are quite sensitive to the choice of the initial wave function $\Psi_i(\mathbf{r_1}, \mathbf{r_2})$ when calculated in the length or velocity form, and less sensitive in the acceleration form [4.61]. Usually, $\Psi_i(\mathbf{r_1}, \mathbf{r_2})$ is presented as expansion on the partial waves [4.59]

$$\Psi_i(\mathbf{r_1}, \mathbf{r_2}) = \frac{1}{4\pi} \sum_l \mathrm{F}_l(r_1, r_2) \mathrm{P}_l(cos\theta_{12}), \tag{4.18}$$

$$\mathrm{F}_l(r_1, r_2) = \sum_{a \geq b} A_{ab}^{(l)} r_1^l r_2^l (r_1^a r_2^b + r_1^b r_2^a) \mathrm{e}^{-B(r_1+r_2)/2}, \tag{4.19}$$

where the sum on a and b runs from 0 to 5 with $a + b \leq 6$ giving 15 terms for each partial wave, $\mathrm{P}_l(x)$ are the Legendre polynomials and θ_{12} is the angle between vectors r_1 and r_2 . The use of the first three terms in (4.18) and the constant B= 3.7 in (4.19) already provide the correct value for the double-ionization potential (I_2 = 79 eV). The expansion coefficients $A_{mn}^{(l)}$ for the ground-state wave function of the He atom are given in Table 4.6.

The initial-state wave function can be also expanded in a *correlated* Hylleraas basis set [4.62, 63]

$$\phi = (1 + P_{12}) r_1^a r_2^b r_{12}^c \mathrm{e}^{(-Ar_1 - Br_2)} Y_{\ell_1 \ell_2}^{LM}(\mathbf{r_1}, \mathbf{r_2}), \; a + b + c \leq 6, \tag{4.20}$$

where $Y_{\ell_1 \ell_2}^{LM}$ are the total angular eigenfunctions corresponding to the states with the orbital L and the magnetic M quantum numbers, and P_{12} is the antisymmetrization operator.

While for the initial wave function Ψ_i one can find quite an accurate expression, the situation is much more complicated for the final wave function Ψ_f which has to describe two electrons in the continuum, and this seems to be one of the main difficulties in the two-photoionization problem [4.64, 65]. In the case of high photon energies, $\hbar\omega \to \infty$, at least one of the ejected electrons takes almost all the photon energy and angular momentum, the other electron having a little energy and zero angular momentum. The correlation between outgoing electrons can be neglected and the fast electron is represented by the plane wave while the other by the Coulomb wave function, so that Ψ_f can be represented as a symmetrized combination:

$$\Psi_f(\mathbf{r_1},\mathbf{r_2}) = \frac{1}{\sqrt{2}}\left[\frac{k_f^{1/2}}{(2\pi)^{3/2}}\right]\left[\mathrm{e}^{\mathrm{i}\mathbf{k}_f\mathbf{r_1}}\varphi(\eta_2,\mathbf{k_2};\mathbf{r_2})+\varphi(\eta_1,\mathbf{k_1};\mathbf{r_1})\mathrm{e}^{\mathrm{i}\mathbf{k_f}\mathbf{r_2}}\right], \tag{4.21}$$

where $\varphi(\eta,\mathbf{k},\mathbf{r})$ are the usual Coulomb wave functions

$$\varphi(\eta,\mathbf{k},\mathbf{r}) = \mathrm{e}^{-\pi\eta/2}\Gamma(1-\mathrm{i}\eta)\mathrm{F}[\mathrm{i}\eta,1;\mathrm{i}(kr+\mathbf{kr})], \tag{4.22}$$

Here $\mathrm{F}(a,b;x)$ is the confluent hypergeometric function, η is the Sommerfeld parameter $\eta = Ze^2/k$ and Z is the effective charge.

A more elaborate (distorted wave) representation of the final wave function is based on the assumption [4.36] that an essentially exact solution of the two-electron Hamiltonian exists when two electrons are separated enough in the phase space, i.e., if the conditions $(k_i,r_i),(k_{12}r_{12}) \gg 1$ are satisfied where r_{12} denotes the interelectronic position. In this representation the final wave function is represented as

$$\Psi_f(\mathbf{r_1},\mathbf{r_2}) = \frac{1}{\sqrt{2}}\left[\frac{k_1^{1/2}}{(2\pi)^{3/2}}\mathrm{e}^{\mathrm{i}\mathbf{k_1}\mathbf{r_1}}\varphi(\eta_2,\mathbf{k_2};\mathbf{r_2})+\frac{k_2^{1/2}}{(2\pi)^{3/2}}\mathrm{e}^{\mathrm{i}\mathbf{k_2}\mathbf{r_2}}\varphi(\eta_1,\mathbf{k_1};\mathbf{r_1})\right] \times\varphi(\eta_{12},\mathbf{k_{12}};\mathbf{r_{12}}). \tag{4.23}$$

The Coulomb distortion factor $\varphi(\eta_{12},\mathbf{k_{12}},\mathbf{r_{12}})$ is given by (4.22) with

$$\mathbf{k}_{12} = \frac{\mathbf{k_1}-\mathbf{k_2}}{2},\quad \eta_{12} = \frac{1}{2k_{12}}. \tag{4.24}$$

and describes the electron–electron interaction. The problems arising in the theory of double photoionization of He are discussed in [4.65].

Theoretical predictions of the double photoionization cross sections $\sigma_2(\omega)$ for He are compared with recommended values [4.14] in Fig. 4.5. The calculations by *Tiwary* [4.68] and *Hino* et al. [4.28] have been normalized

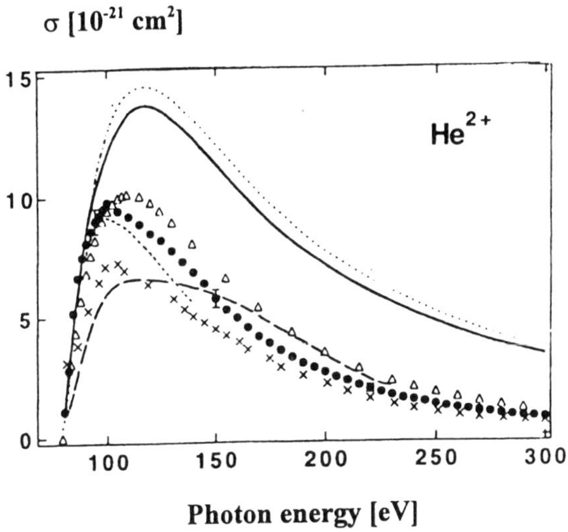

Fig. 4.5. Double photoionization cross section of He. Experiment: solid circles, recommended data [4.14]. Theory: dotted curve, multiconfiguration self-consistent field [4.66]; solid curve, initial-state configuration-interaction [4.67]; open triangles, MBPT, L-form [4.27]; short-dashed curve, quasiclassical final state [4.35]; long-dashed curve, initial-state configuration-interaction [4.68]; crosses, MBPT [4.28]. From [4.14]

to the total photoabsorption cross sections measured in [4.14]. In general, there is quite large discrepancy in results of different calculations related ith the use of different forms of the applied dipole operator and with the spread of the experimental $\sigma_2(\omega)/\sigma_{\text{tot}}(\omega)$ ratios. The best agreement with the recommended data is obtained by the MBPT (L-form) calculations of *Carter* and *Kelly* [4.27].

Cross section calculations by *Kornberg* and *Miraglia* [4.69] with the correlated continuum wave functions are 3 times higher than experimental data. Variational calculations of *Nicolaides* et al. [4.40] and R-matrix results of *Meyer* et al. [4.33] are also not in a good agreement with the recommended experimental data. (These theoretical data are not shown in Fig. 4.5).

The dependence of the experimental and theoretical ratios $R_2(\omega)$ on photon frequency for He is shown in Fig. 4.6. The ratio has its maximum (about 4.8%) at 200 eV and then reaches a constant value (about 1.7%) at photon enegies near 4 keV. The limiting ratio has been estimated to be 1.6–2.3% [4.59, 73–77] and now is adopted as 1.66% [4.30, 37, 78]. The experimental data in Fig. 4.6 are in good agreement with MBPT calculations of *Hino* et al. [4.28] in the L, V and A forms. The difference between the A form and V (or L) form near the $R_2(\omega)$ maximum is negligible at high energies.

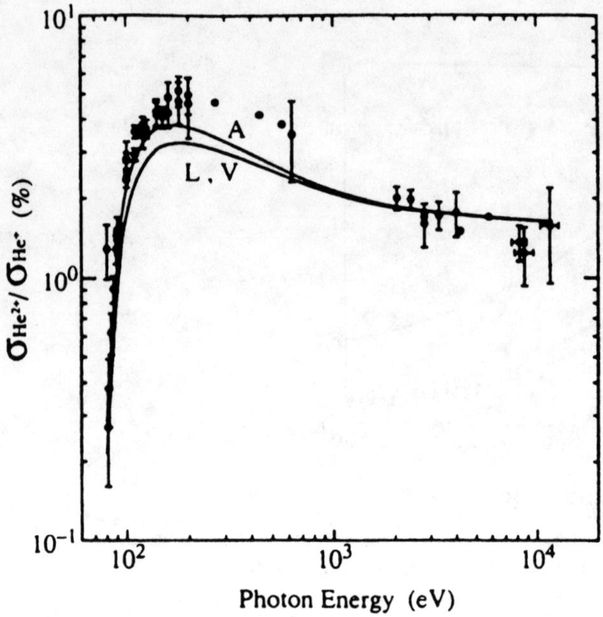

Fig. 4.6. Cross-section ratios $R_2(\omega)$ for double- and single-photoionization of He. Symbols: experimental data from [4.10,11,53,55–57,70–72]. Solid curves: the L, V and A forms of MBPT calculations of *Hino* et al. [4.28]. From [4.28]

By this time, the most accurate measurements of $R_2(\omega)$ for He at energies $\hbar\omega > 2$ keV have been carried out in the works [4.79,80]. But at higher energies, $\hbar\omega \geq 6$ keV, the ratio $R_2(\omega)$ to create the He atom in doubly or singly ionized states, is not independent anymore of frequency ω because of the contribution of the Compton scattering, (4.2) (Sect. 4.2.2).

DPI cross section of H^- ions ($I_2 = 14.3$ eV) has been measured [4.21] near threshold region $\hbar\omega \leq 14.65$ eV but these values refer to the *relative* cross sections. A behavior of MPI cross sections of atoms and ions near threshold constitutes a special problem, so-called Wannier theory [4.81–83]. According to classical and quantum-mechanical approaches, the DPI cross section $\sigma_2(\omega)$ is described by a power law in the threshold energy range

$$\sigma_2(\omega) \approx A(\hbar\omega - I_2)^a + B, \quad \hbar\omega \approx I_2, \tag{4.25}$$

where a, A and B are constants. For the H^- ion the Wannier theory predicts $a = 1.127$. The measured cross section $\sigma_2(\omega)$ for H^- is depicted in Fig. 4.7 in comparison with theoretical predictions. The values of a, A and B derived from experimental data are

$$a = 1.15 \pm 0.04, \quad A = 38.5 \pm 1.5, \quad B = 0.68 \pm 0.05\,. \tag{4.26}$$

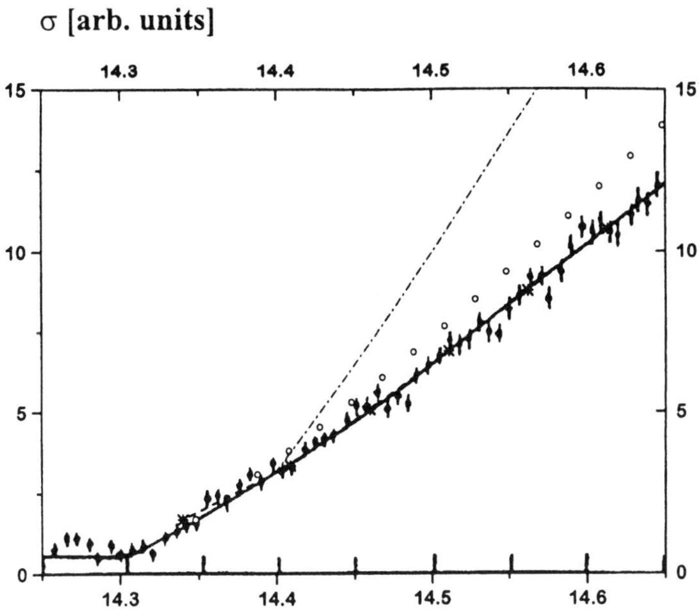

Fig. 4.7. Double-photoionization cross section of H^- (in arbitrary units) near threshold. Solid circles – experiment [4.21]; solid line – fitting by (4.25, 26). Dash-dotted curve – calculated value [4.84] in the L-form equated to experimental data at $\hbar\omega = 14.4$ eV. The open circles and stars – variational calculations [4.40] in the $L-$ and $V-$ forms, respectively, equated to experimental data at $\hbar\omega =$ 14.355 eV. From [4.40]

Theoretical calculations shown in Fig. 4.7 are equated with experimental data [4.21] at different photon energies (see figure caption).

Calculated cross sections $\sigma_2(\omega)$ for H^- for photon energies up to ≈ 100 eV are displayed in Fig. 4.8. There is a significant difference between the calculated cross sections in the L and V forms.

At the high-ω limit, some results can be obtained in a close analytical form. Using expression (4.21) and the method of *Kabir* and *Salpeter* [4.85], it was shown [4.59] that as $\hbar\omega \to \infty$ the single and the total photoionization cross sections can be expressed as

$$\sigma_1(\omega \to \infty) = \frac{256\pi^2 \alpha a_0^2}{3\sqrt{2}(\hbar\omega)^{7/2}} \sum_B \mid \int \Psi_i(\mathbf{r_1}, 0)\phi_B^*(\mathbf{r_1}) d\mathbf{r_1} \mid^2, \tag{4.27}$$

and

$$\sigma_{\text{tot}}(\omega \to \infty) = \frac{256\pi^2 \alpha a_0^2}{3\sqrt{2}(\hbar\omega)^{7/2}} \int \mid \Psi_i(\mathbf{r_1}, 0) \mid^2 d\mathbf{r_1}\,, \tag{4.28}$$

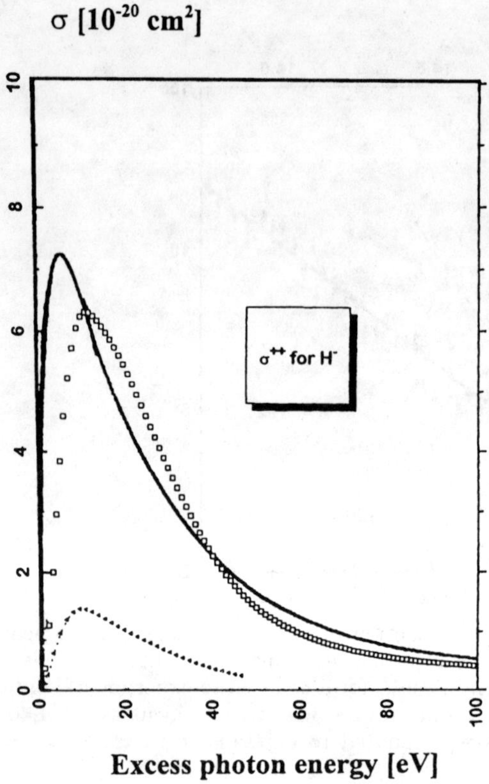

Fig. 4.8. Calculated double-photoionization cross section of H^- as a function of excess photon energy. Solid curve – the L-form calculations in the initial-state configuration interaction approach [4.67]. Open squares and triangles – the L and V forms of variational approach [4.40]. From [4.40]

respectively. Equation (4.27) corresponds to the cross section for one-electron photoionization with simultaneous excitation of the H-like core (He^+) and (4.28) to all possible photoionization processes (both single and double).

The most striking features followed from (4.27, 28) are:

(i) asymptotic behavior of single and total photoionization cross sections depends only on the behavior of the *initial* ground-state wave function, but not on the *final* wave function. In other words, in this simple asymptotic analysis, the most significant aspect of the problem – the electron–electron correlation in the final state – is ignored;

(ii) both cross sections decrease with the photon frequency by a similar law, i.e., as $\omega^{-7/2}$. This energy dependence of the total photoionization in

He has been recently measured [4.70, 86] at energies greater than about 2.4 keV [4.26].

From these properties it follows that for two-electron systems the ratio of cross sections for double- to single-photoionization, $R_2^{\rm ph}(\omega)$, approaches the constant at a high photon energy

$$R_2^{\rm ph}(\infty) = 1 - \frac{\sum_B \mid \int \Psi_i(\mathbf{r_1}, 0)\phi_B^*(\mathbf{r_1})d\mathbf{r_1} \mid^2}{\int \mid \Psi_i(\mathbf{r_1}, 0) \mid^2 d\mathbf{r_1}} . \tag{4.29}$$

Equation (4.29) is believed to be correct for all nuclear charges Z of He-like systems. The Z-dependence of the photoeffect ratio $R_2^{\rm ph}(\infty)$ for He-like systems has been studied in [4.38, 78]. For high-Z ions the result reads

$$R_2^{\rm ph}(\infty) \approx \frac{0.09}{Z^2} - \frac{0.03}{Z^3}, \quad Z \gg 1, \tag{4.30}$$

which is in close agreement with earlier prediction by *Amusia* et al. [4.76]

$$R_2^{\rm ph}(\infty) \approx \frac{0.093}{Z^2}, \quad Z \gg 1 . \tag{4.31}$$

The calculated asymptotically exact ratios $R_2^{\rm ph}(\infty)$ [4.30, 87] for He-like systems (H^-, He, Li^+) and molecular hydrogen (H_2) are given in Table 4.7; these values are in good agreement with other sophisticated calculations.

Table 4.7. Asymptotic raios $R_2^{\rm ph}(\infty)$ of photoeffect cross sections for two-electron systems. From [4.30, 87]

System	$R_2^{\rm ph}(\infty)$, %
H^-	1.50
He	1.66
Li^+	0.87
H_2	2.25

Double-photoionization cross sections of atoms and ions define the logarithmic (Bethe) constants of high-energy double-ionization cross sections $\sigma_2(E)$ by a charged particle. According to [4.59], at high electron energies the cross section $\sigma_2(E)$ can be presented in the form

$$\sigma_2(E) \approx \frac{2\pi a_0^2}{E}\left(A\ln E + B + \frac{C}{E}\right), \quad E \gg I_2, \tag{4.32}$$

where the logarithmic constant A can be found through the known double-photoionization cross section

$$A = (4\pi\alpha a_0^2)^{-1} \int_{I_2}^{\infty} \sigma_2(\omega)\frac{d\hbar\omega}{\hbar\omega} . \tag{4.33}$$

This is the so-called Byron–Joachain relation between $\sigma_2(E)$ and $\sigma_2(\omega)$ which can be generalized for m-electron ionization [4.26]. Equations (4.32, 33) are in accordance with results obtained by the more general Weizsäcker–Williams method of equivalent photons (Sect. 4.1).

According to [4.38], the double-photoionization cross sections $\sigma_2(\omega)$ for He-like systems are scaled as

$$Z^4\sigma_2(\omega) = F(E_f/Z^2), \ E_f = \hbar\omega - I_2, \tag{4.34}$$

where Z is the nuclear charge of the target, E_f is the total kinetic energy of two free electrons (excess photon energy) and $F(x)$ is a certain function. In the work [4.88] it was found that the use of the screened (effective) nuclear charge

$$Z_{eff} = Z - \frac{5}{16}, \tag{4.35}$$

instead of Z, leads to a better description of the double-ionization cross sections of low-Z systems (H^-) by electron impact using (4.32, 33) (Sect. 2.1).

4.2.2 Compton Scattering

As was mentioned in Sect. 4.1, at high photon energies the Compton scattering

$$\hbar\omega_i + \mathrm{A} \rightarrow \hbar\omega_f + \mathrm{A}^{\mathrm{m}+} + m\mathrm{e}^-, \ \ m \geq 1, \tag{4.36}$$

begins to play the main role as schematically shown in Fig. 4.9 for one- and two-electron photoionization cross sections of He. As it has been seen, at photon energies higher than about 6 keV, it is quite difficult to distinguish between photoeffect and Compton scattering. However, the recent experimental techniques is able to separate these two photoionization mechanisms [4.70, 90, 91].

In the case of He, the ratio $R_2^{\mathrm{ph}}(\omega)$ of double- to single-ionization cross sections including both photoeffect and Compton contributions reads

$$R_2(\omega) = \frac{\sigma_2^{\mathrm{ph}}(\omega) + \sigma_2^{\mathrm{c}}(\omega)}{\sigma_1^{\mathrm{ph}}(\omega) + \sigma_1^{\mathrm{c}}(\omega)}. \tag{4.37}$$

Here indices ph and c refer to the pure photoeffect and Compton scattering, respectively. The main differences between photoeffect and Compton scattering are [4.29, 46]:

i) in Compton scattering, the relevant interaction operator is (in the first order) $\mathbf{A}^2$ rather than $\mathbf{p}\cdot\mathbf{A}$ for the photoeffect,

ii) the most probable energies for electrons ejected in Compton scattering are significantly smaller than that of the photoionized electrons,

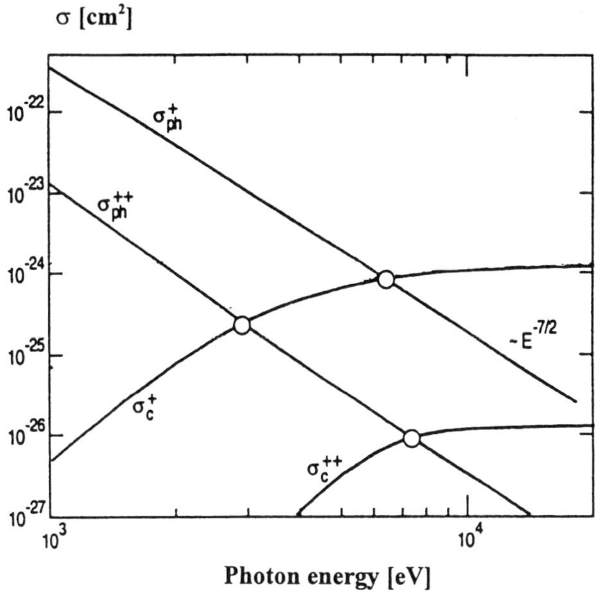

Fig. 4.9. Calculated single σ^+ and double σ^{++} cross sections of photoeffect (ph) and Compton scattering (c) as a function of photon energy E. From [4.89]

iii) for Compton scattering, the dipole approximation used in calculations of photoionization, is not valid anymore because the momentum transfer of photon to an atom is relatively large, and more terms of the multipole expansion of the photon fields should be taken into account.

The differential cross section for double ionization within the $\mathbf{A^2}$ approximation is given by [4.29]

$$\frac{d^2\sigma_2^c}{d\omega d\Omega} = \left(\frac{d\sigma}{d\Omega}\right)_{Th} \left(\frac{\omega_f}{\omega_i}\right) \int d\mathbf{p_1} d\mathbf{p_2} \mid \langle \Psi_f \mid \sum_{j=1,2} \mathrm{e}^{\mathrm{i}\mathbf{k}\mathbf{r}_j} \mid \Psi_i \rangle \mid^2$$
$$\times \delta(E_f - E_i - \hbar\omega), \tag{4.38}$$

where $(d\sigma/d\Omega)_{Th}$ is the classical Thompson scattering cross section, Ω is the solid angle of the scattered photon, $\Psi_{\mathrm{i,f}}$ and $E_{\mathrm{i,f}}$ are the initial and final wave functions and energies of the target, and $p_{1,2}$ are momenta of two ejected electrons. One can see that the Compton photoionization cross section is expressed through the generalized oscillator strength (matrix elements of $\mathrm{e}^{\mathrm{i}\mathbf{k}\mathbf{r}}$).

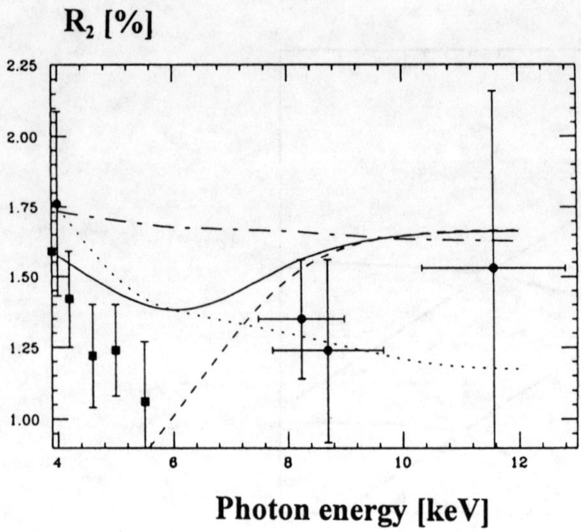

Fig. 4.10. The ratio R_2 of double- to single-ionization cross sections of He at high photon energies. Experiment: full circles – [4.56], squares – [4.93]. Theory: chain-dashed curve – $R_2^{\rm ph}$ ratio for photoeffect, dashed curve – $R_2^{\rm c}$ for Compton scattering and solid curve – the ratio incorporating both the contributions (MBPT calculations [4.29]). The dotted curve – estimation [4.37] of the total R_2 ratio. From [4.29]

A high-ω limit formula of the double-to-total ratio $R_2^{\rm c}(\infty)$ for Compton scattering is given by [4.92]

$$R_2^{\rm c}(\infty) = 1 - \sum_B \int d\mathbf{r} \mid \int \Psi_i(\mathbf{r_1}, \mathbf{r_2}) \phi_B^*(\mathbf{r_1}) d\mathbf{r_1} \mid^2, \tag{4.39}$$

where the summation is over all the bound states B of the hydrogen-like wave functions ϕ_B of the residual ion.

Individual ratios $R_2(\omega)$ due to photoeffect and Compton scattering and the total $R_2(\omega)$ value defined by (4.37) are shown in Fig. 4.10 in comparison with experimental data. It has been seen that the MBPT calculations [4.29] display a shallow dip structure around 6 keV which Hino et al. [4.29] explained as a transitional change of the dominant ionizing process from photoeffect to Compton scattering. At photon energies $\hbar\omega \geq 10$ keV, the Compton ratio is approaching an asymptotic value of 1.67% which is practically the same as predicted for the pure photoeffect that is a surprising coincidence [4.70, 94].

Calculations of the Compton ratio $R_2^{\rm c}(\omega)$ have been performed in [4.29, 64, 92, 95–97]. Although for finite photon energies ($\hbar\omega \leq$20 keV) these calculations disagree significantly, in the high-ω limit various calculations approach the same constant for the He atom: $R_2^c(\infty) = 0.08$.

The Z-dependence of $R_2^c(\infty)$ has been investigated in [4.46, 95, 96, 98]. According to results by *Amusia* and *Mikhailov* [4.95, 96]

$$R_2^c(\infty) \approx \frac{0.048}{Z^2}, \quad Z \gg 1, \tag{4.40}$$

which is about a half of the corresponding results for the photoeffect ratio $R_2^{ph}(\infty)$ given in (4.31).

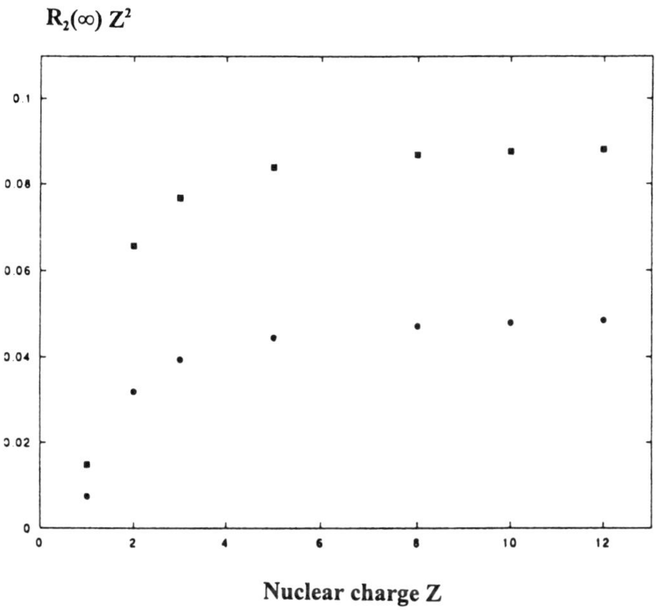

Fig. 4.11. Calculated $R_2^c(\infty)Z^2$ (circles) and $R_2^{ph}(\infty)Z^2$ (squares) as a function of the nuclear charge Z of He-like systems. In the high-Z limit, $R_2^c(\infty)Z^2 \sim 0.05$ and $R_2^c(\infty)Z^2 \sim 0.09$. From [4.98]

Figure 4.11 shows calculations of $R_2^{ph}(\infty)$ and $R_2^c(\infty)$ [4.98] as a function of the nuclear charge Z for He-like systems from H^- up to Mg^{10+} using the Hylleraas-type initial-state wave functions. The results for $R_2^{ph}(\infty)$ are in close agreement with those by *Forrey* et al. [4.78] , and for high Z they are in agreement with (4.31). As for the Compton ratio, these results are in a good agreement with (4.40) for $Z \gg 1$. In general, the $R_2^c(\infty)$-values are about 2 times smaller than $R_2^{ph}(\infty)$.

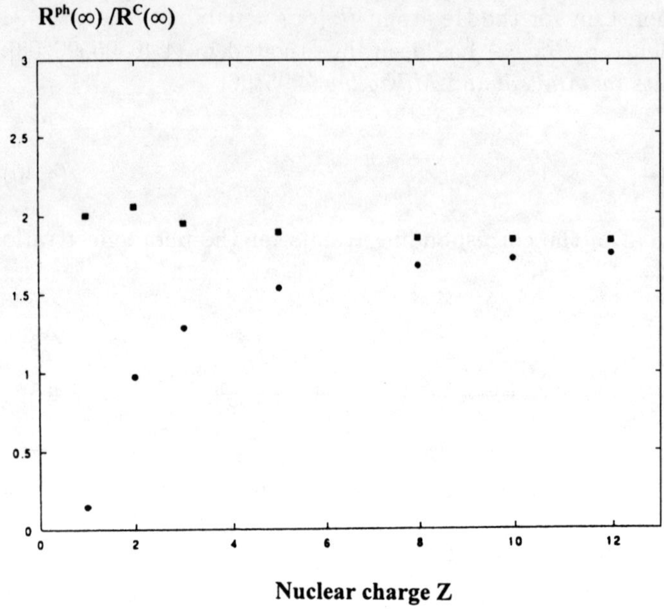

Fig. 4.12. Calculated Z-dependence of the ratio $R_2^{\mathrm{ph}}(\infty)/R_2^{\mathrm{c}}(\infty)$ for He-like ions: circles – [4.95, 96], squares – [4.98]. From [4.98]

The calculated ratio $R = R_2^{\mathrm{ph}}(\infty)/R_2^{\mathrm{c}}(\infty)$ as a function of Z for He-like ions is shown in Fig. 4.12. Calculations of *Amusia* and *Mikhailov* [4.95, 96] show a smooth dependence of R on Z, and, in fact, they give $R = 1$ for the He atom. According to [4.98], the ratio $R \approx 2$ for all Z. With this ratio $R \approx 2$, Surić et al. [4.98] have connected the single- and double-photoionization cross sections for all He-like systems by a simple expression

$$\frac{\sigma_2^{\mathrm{c}}(\omega)}{\sigma_2^{\mathrm{ph}}(\omega)} \approx \frac{1}{2}\frac{\sigma_1^{\mathrm{c}}(\omega)}{\sigma_1^{\mathrm{ph}}(\omega)}, \quad \omega \to \infty . \tag{4.41}$$

4.3 Photoionization with Excitation

Similar to the ionization with excitation of atoms and ions by electron impact, considered in Sect. 2.4, simultaneous photoionization and excitation of the core electron induced by a single high-energy photon impact can occur. For example, photoionization of the inner $3p$-electron in atomic potassium can occur with simultaneous excitation of the outer $4s$-electron

$$\hbar\omega + \mathrm{K}(3p^6 4s\ ^2S) \to \mathrm{K}^+(3p^5 ns\ ^{1,3}P) + \mathrm{e}^-(\epsilon s, \epsilon d), \ n \geq 4, \tag{4.42}$$

Table 4.8. The values of the overlap integrals $< n'l' \mid n''l' >$ between radial orbitals of the configurations $nl^{4l+2}n'l'$ and $nl^{4l+1}n''l'$ for Li, Na and K. From [4.118]

Atom	Term LS	$< n'l' \mid n''l' >$	Integral	$< n'l' \mid n''l' >$	Integral
Li	^1S	$< 2s \mid 2s >$	0.9102	$< 2s \mid 3s >$	-0.3972
	^3S	$< 2s \mid 2s >$	0.8724	$< 2s \mid 3s >$	-0.4666
	^1P	$< 2p \mid 2p >$	0.7595	$< 2p \mid 3p >$	-0.5879
	^3P	$< 2p \mid 2p >$	0.7595	$< 2p \mid 3p >$	-0.6462
Na	^1P	$< 3s \mid 3s >$	0.9275	$< 3s \mid 4s >$	-0.3569
	^3P	$< 3s \mid 3s >$	0.9146	$< 3s \mid 4s >$	-0.3846
	^1S	$< 3p \mid 3p >$	0.8347	$< 3p \mid 4p >$	-0.5443
	^3S	$< 3p \mid 3p >$	0.808	$< 3p \mid 4p >$	-0.5920
	^1P	$< 3p \mid 3p >$	0.8407	$< 3p \mid 4p >$	-0.5349
	^3P	$< 3p \mid 3p >$	0.8407	$< 3p \mid 4p >$	-0.5349
	^1D	$< 3p \mid 3p >$	0.8387	$< 3p \mid 4p >$	-0.5385
	^3D	$< 3p \mid 3p >$	0.8254	$< 3p \mid 4p >$	-0.5578
K	^1P	$< 4s \mid 4s >$	0.9414	$< 4s \mid 5s >$	-0.3230
	^3P	$< 4s \mid 4s >$	0.9479	$< 4s \mid 5s >$	-0.3545
	^1S	$< 4p \mid 4p >$	0.8706	$< 4p \mid 5p >$	-0.4845
	^3S	$< 4p \mid 4p >$	0.8330	$< 4p \mid 5p >$	-0.5451
	^1P	$< 4p \mid 4p >$	0.8798	$< 4p \mid 5p >$	-0.4676
	^3P	$< 4p \mid 4p >$	0.8798	$< 4p \mid 5p >$	-0.4676
	^1D	$< 4p \mid 4p >$	0.8749	$< 4p \mid 5p >$	-0.4774
	^3D	$< 4p \mid 4p >$	0.8619	$< 4p \mid 5p >$	-0.4989

where ϵ is the energy of the photoelectron. Such a reaction is a double-electron process when under photon impact one target electron is ionized and another one is excited.

If the K^+ ion is created in the state with the principal quantum number n=4, the process (4.42) can be described by the so-called *frozen core* approximation when the wave functions of the initial and final states are calculated in the same field of the core. However, the single photoionization can occur with simultaneous excitation (sometimes also called *shake-off* process) of one or more electrons when the atomic core behaves as an unfrozen or relaxed system.

Such double-electron transitions are of a special interest. A creation of inner-shell vacancy leads to a strong rearrangement of the atomic core which has to be accounted for calculation of the wave functions in the final (continuum) states. The effects of the unfrozen core are very important for characteristics of many-electron atoms making multielectron transitions such as autoionization rates of core excited atoms [4.99–101], K_α-transition probabilities [4.102, 103], transitions between d- and f-subshells in heavy neutral atoms where the collapse of the radial orbitals is very strong. Sophisticated methods describing account of the core relaxation effects (many-body perturbation theory, close coupling, R-matrix, configuration interaction) are described in [4.20, 104].

Simultaneous photoionization and excitation processes of alkali and rare-gas atoms into the excited states have been considered experimentally [4.14, 105–112] and theoretically [4.113–119].

In the simple approximation, the shake-off processes (excitation of the core electron) are described by products of the overlap integrals of the radial electron wave functions. For example, the photoionization cross section for transition $i \to f$ with excitation can be written in the form [4.118]

$$\sigma(i \to f) = 4\pi^2 \alpha a_0^2 \frac{\hbar\omega}{3 g_i Ry} \sum_{\lambda} \sum_{L'S'} |< i \mid \mathbf{D} \mid f\epsilon\lambda L'S' >|^2, \tag{4.43}$$

where ω is the photon frequency, α is the fine-structure constant, a_0 is the Bohr radius, g_i is the statistical weight of the initial state, ϵ and λ are the energy and angular momentum of the photoelectron, $L'S'$ are the total angular momenta in the final state and $\mathbf{D}$ is the operator of electric dipole radiation in the length or velocity gauge. For transition $i = K_i n_i l^{4l+2} n'l'LS \to f = K_f n_f l^{4l+1} n''l'' L_1 S_1 \epsilon\lambda L'S'$ (K_i and K_f are the closed shells), one has:

$$\begin{aligned} &< i \mid \mathbf{D} \mid f\epsilon\lambda L'S' >= \delta(S,s)\delta(L,l') < n_i l \mid n_f l >^{4l+1} \\ &\times \prod_{n_i l_i} < n_i \mid n_f >^{4l_i+2} \delta(n_i n_f)\delta(l_i l_f) \\ &\times \{< l \parallel C^{(1)} \parallel \lambda >< n_i l \mid R \mid \epsilon\lambda >< n'l' \mid n''l'' > Q^{(1)} \\ &+ < l \parallel C^{(1)} \parallel l' >< n_i l \mid R \mid n''l'' >< n'l' \mid \epsilon\lambda > Q^{(2)} \\ &+ < l' \parallel C^{(1)} \parallel \lambda >< n'l' \mid R \mid \epsilon\lambda >< n_i l \mid n''l'' > Q^{(3)}\} \end{aligned} \tag{4.44}$$

where $<\parallel C^{(1)} \parallel>$ and $<\mid R \mid>$ are the reduced and dipole matrix elements, respectively and Q are the angular coefficients depending on the angular and spin momenta of the initial and final states of the system.

The first term in (4.44) describes photoionization of the nl-electron from the closed subshell and excitation of the outermost electron. The second and third terms give the contribution of the photoionization of an electron from the open shell and the excitation of an electron from the inner shell, respectively.

Figure 4.13 shows cross sections of photoionization with excitation of the He atom with He^+ being left in the excited $n = 2$ state. At photon energies below 100 eV, all calculations overestimate the experimental recommended cross section [4.14]. At higher energies, all data are in agreement with each other.

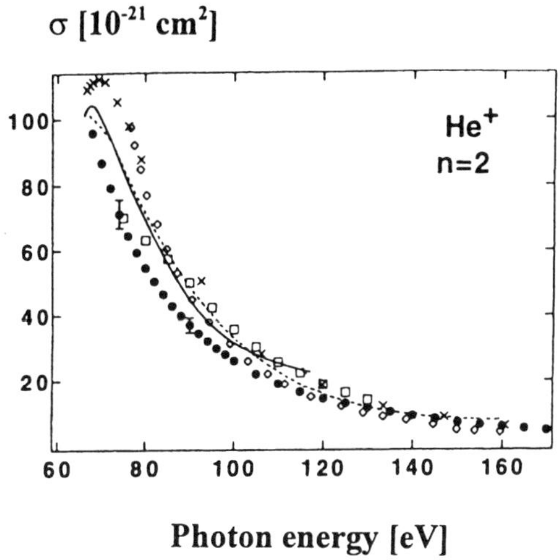

Fig. 4.13. Photoionization cross sections of He with simultaneous excitation of the He^+ ion into the $n = 2$ state. Solid circles: recommended experimental data [4.14]. Theories: crosses – close-coupling calculations [4.120]; open diamonds – the relaxed Hartree-Fock method [4.121]; open squares – MBPT [4.122]; solid curve – R-matrix calculations [4.123]; dashed curve – MBPT [4.124]. From [4.14]

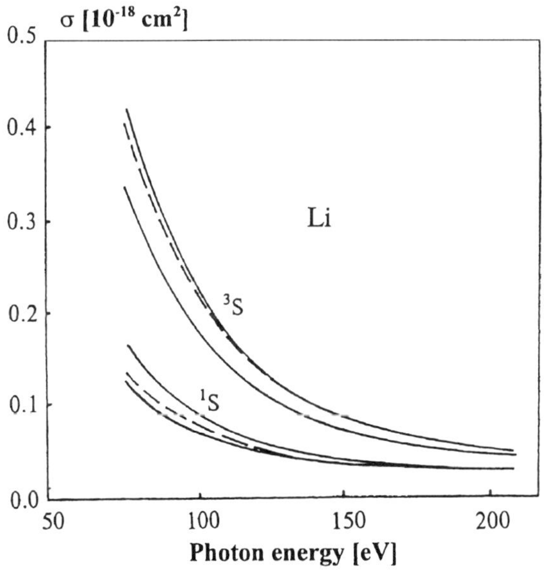

Fig. 4.14. Calculated cross sections [4.118] for reactions $1s^2 2s\ ^2S - 1s3s\ ^1S,\ ^3S$ in Li, i.e., for simultaneous photoionization of the inner $1s$-electron and excitation of the onter $2s$-electron into $3s$-state, (4.44). Solids curves – length gauge, dashed curves – velocity gauge. Lower solid curves correspond to calculations of the radial integrals for specific terms

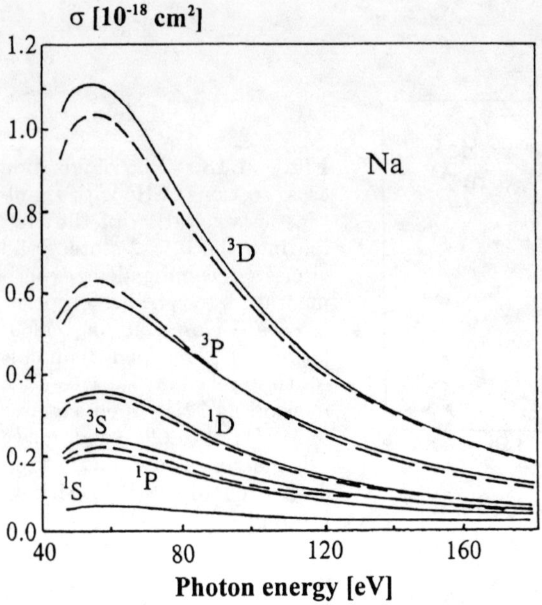

Fig. 4.15. Calculated cross sections for reactions $2p^6 3p\ ^2P - 2p^5 4p\, LS\epsilon d$ in Na, i.e., for simultaneous $2p \to \epsilon d$ photoionization and $3p - 4p$ excitation, (4.44); LS terms are indicated. Solids curves correspond to calculations of specific terms and dashed curves for the average configurations. From [4.118]

Absolute and relative photoionization cross sections with excitation for alkali atoms are depicted in Figs. 4.14–17. The overlap integrals of the radial orbitals of the outermost electron in the initial and final states strongly influence the photoionization cross sections, seen from (4.44). The transition integrals are also very sensitive to the atomic states. The numerical values of the overlap integrals $< n'l' \mid n''l' >$ are presented in Table 4.8. If the relaxation effects would be neglected, the overlap integrals $< n'l' \mid (n' + 1)l' >$ equal zero.

According to [4.118,119], photoionization with excitation is more important for excited states as compared to the ground state. An enhancement of relative intensities is caused by the rearrangement of electrons in the final state that are further from the nucleus than those involved in the process. A very high sensitivity of the radial orbitals of electron in the continuum to the approximation used for calculations in the case of Li and K atoms as seen from Figs. 4.14–17. To obtain reliable cross sections for the photon energies near threshold, it is necessary to perform calculations with the radial orbitals

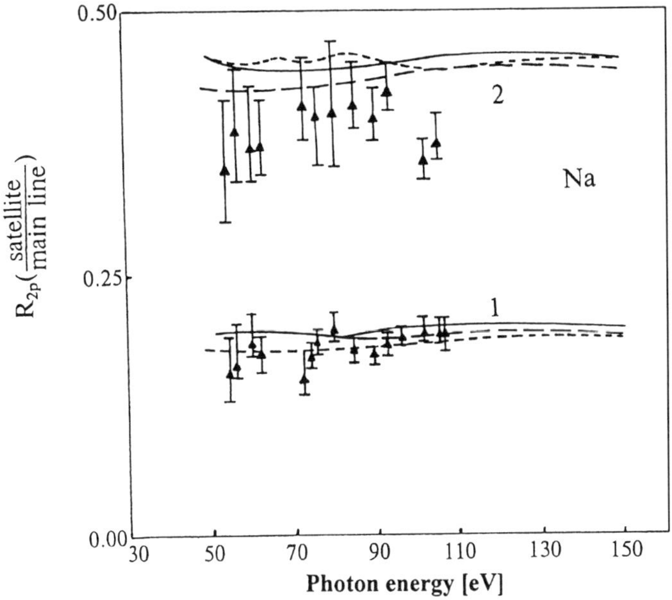

Fig. 4.16. Ratios of the total photoionization cross section with excitation to the pure photoionization cross section of the inner $2p$-electron in Na. Set 1 is for Na in the ground state, and set 2 for excited $2p^6 3p\ ^2P$ term. Solid curves – calculation in the velocity gauge, and dotted curves – calculation with the radial integrals for specific terms; triangles – experiment [4.108]. From [4.118]

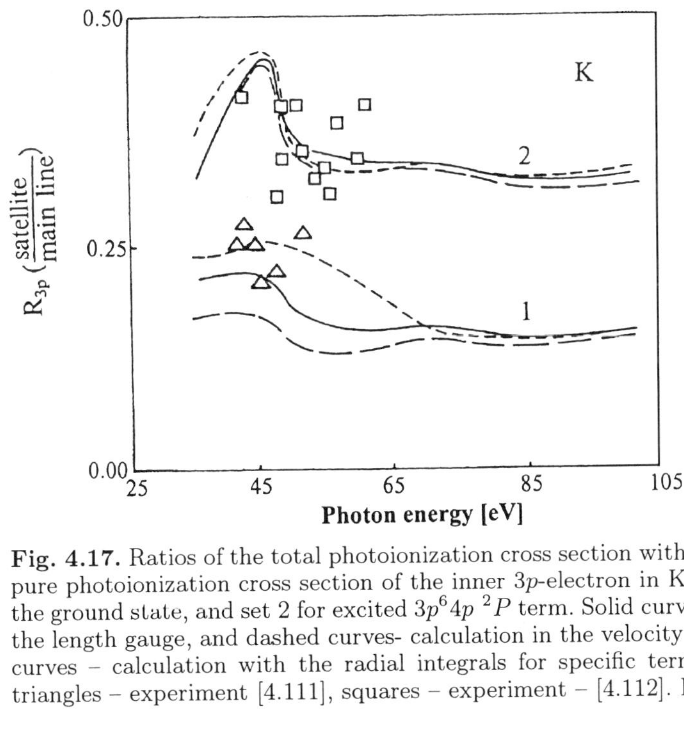

Fig. 4.17. Ratios of the total photoionization cross section with excitation, to the pure photoionization cross section of the inner $3p$-electron in K. Set 1 is for K in the ground state, and set 2 for excited $3p^6 4p\ ^2P$ term. Solid curves – calculation in the length gauge, and dashed curves- calculation in the velocity gauge and dotted curves – calculation with the radial integrals for specific terms (from [4.118]); triangles – experiment [4.111], squares – experiment – [4.112]. From [4.118]

of specific terms. Differences between the cross sections calculated with the radial orbitals of the specific term and that of the averaged configurations are much larger than those between the cross sections of the length and velocity forms. The cross sections for Na atoms are insensitive to the approximation used for the radial orbitals.

References

Chapter 1

[1.1] L. Valyi: *Atom and Ion Sources* (Wiley, New York 1977)
[1.2] G. Scoles: *Atomic and Molecular Beam Methods* (Oxford Univ. Press, Oxford 1988)
[1.3] S. Trajmar, J.C. Nickel: Adv. At. Mol. Opt. Phys. **30**, 45 (1992)
N. Andersen: *Atomic Collisions with Laser-Excited Targets*: in: *Review of Fundamental Processes and Application of Atoms and Ions*, ed. by C.D. Lin (World Scientific, Singapore 1993) p. 213
[1.4] H. Tawara: Unpublished (1986)
[1.5] J. Ullrich, R. Dörner, V. Mergel, O. Jagutzki, L. Spielberger, H. Schmidt-Böcking: Commun. At. Mol. Phys. **30**, 285 (1994)
[1.6] M. Gieler, F. Aumayr, J. Schweinzer, W. Koppensteiner, W. Husinsky, H.-P. Winter, K. Lozkin, J.P. Hansen: J. Phys. B **26**, 2137 (1993)
[1.7] B.D. DePaola, M.T. Huang, S. Winecki, Y. Kanai, S.R. Lundeen, C.W. Fehrenbach, S.A. Arko: Phys. Rev. A **52**, 2136 (1995)
[1.8] B.D. DePaola: In *Accelerator-Based Rydberg Atom Colision Experiments in Accelerator-Based Atomic Physics Techniques and Applications*, ed. by S. Shafroth and J.C. Austin (AIP, New York 1997)
[1.9] J.C. Houver, D. Dowek, C. Richter, N. Andersen: Phys. Rev. Lett. **68**, 162 (1992)
[1.10] C. Richter, N. Andersen, J.C. Brenoit, D. Dowek, J.C. Houver, J. Salgado, J.W. Thomsen: J. Phys. B **26**, 723 (1993)
[1.11] M.B. Shah, D.S. Elliott, H.B. Gilbody: J. Phys. B **20**, 3501 (1987)
[1.12] A. Müller, K. Tinschert, C. Achenbach, E. Salzborn, R. Becker: Nucl. Instrum. Methods B **10**, 204 (1985)
[1.13] M. Stenke, K. Aichele, D. Hathiramani, G. Hofmann, M. Steidl, R. Völpel, E. Salzborn: Nucl. Instrum. Methods B **98**, 573 (1995)
[1.14] K. Shima, N. Kuno, M. Yamanouchi, H. Tawara: At. Data and Nucl. Data Tables **51**, 173 (1992)
[1.15] H.D. Betz: Rev. Mod. Phys. **44**, 465 (1973)
[1.16] R. Geller, B. Jacquot: Phys. Scr. T **3**, 19 (1983);
J. Arianer, R. Geller: Ann. Rev. Nucl. Sci. **31**, 19 (1981);
Y. Yongen, C.M. Lyneis: Electron cyclotron resonance ion sources, in *The Physics and Technology of Ion Sources*, ed. by I.G. Brown (Wiley, New York 1989) p. 207
[1.17] E.D. Donets: Electron beam ion source, in *The Physics and Technology of Ion Sources*, ed. by I.G. Brown (Wiley, New York 1989) p. 245

[1.18] M.P. Stöckli: The Operation of electron beam ion sources for atomic physics, in *Accelerator Based Atomic Physics Techniques and Applications*, ed. by S. Shafroth and J.C. Austin (AIP, New York 1997)
[1.19] M.A. Levine, R.E. Marrs, J.R. Henderson, D.A. Knapp, M.B. Schneider: Phys. Scr. T **22**, 157 (1988)
[1.20] D.A. Knapp, R.E. Marrs, S.R. Elliott, E.W. Magee, R. Zasadzinski: Nucl. Instrum. Methods A **334**, 305 (1993)
[1.21] D. Schneider, D.A. Church, G. Weinberg, J. Steiger, B. Beck, J. McDonald, E. Magee, D. Knapp: Rev. Sci. Instr. **65**, 3472 (1994)
[1.22] J. Moseley, W. Aberth, J.R. Peterson: Phys. Rev. Lett. **9**, 435 (1970)
[1.23] K. Dolder, B. Peart: Rep. Prog. Phys. **48**, 1283 (1985)
[1.24] S. Szücs, M. Karemera, M. Terao, F. Brouillard: J. Phys. B **17**, 1613 (1984)
[1.25] C.C. Harvener, A. Müller, P.A. Zeijlmans van Emmichoven, R.A. Phaneuf: Phys. Rev. A **51**, 2982 (1995)
[1.26] K. Okuno: J. Phys. Soc. Jpn. **5**, 1504 (1986)
[1.27] H.-J. Kluge: Nucl. Instrum. Methods B **98**, 500 (1995)
[1.28] H. Beyer, H.-J. Kluge, V.P. Shevelko: *X-Ray Radiation of Highly Charged Ions*, Springer Series Atoms and Plasmas, Vol. 19 (Springer, Berlin, Heidelberg 1997)
[1.29] P.H. Mokler, Th. Stöhlker, C. Kozhuharov, R. Moshammer, P. Rymuza, F. Bosch, T. Kandler: Phys. Scr. T **51**, 28 (1994)
[1.30] B. Franzke: Phys. Scr. T **22**, 41 (1988)
[1.31] P.H. Mokler: AIP Proc. **274**, 515 (1993)
[1.32] GSI-Nachrichten 4/96, 8 (Darmstadt, 1996)
[1.33] C.J. Herrlander, L. Bragges, A. Bárány, H. Danared, P. Heikkinen, S. Hultberg, L. Liljeby, Th. Lindblad: IEEE Trans. NS-**32**, 2718 (1987)
[1.34] R. Stensgaard: Phys. Scr. T **22**, 315 (1988)
[1.35] T. Tanabe, K. Noda, T. Honma, M. Kodaira, K. Chiba, T. Watanabe, A. Noda, S. Watanabe, A. Mizobuchi, M. Yoshizawa, T. Katayama, H. Muto: Nucl. Instrum. Methods A **307**, 7 (1991)
[1.36] A. Müller: Comm. At. Mol. Opt. Phys. **32**, 143 (1996)
[1.37] S. Martin, A. Denis, Y. Querdane, M. Care: Phys. Lett. A **165**, 441 (1992)
[1.38] S. Martin, A. Denis, A. Delon, J. Desesquelles, Y. Querdane: Phys. Rev. A **48**, 1171 (1993)
[1.39] P. Roncin, M.N. Gaboriaud, Z. Szilagy, M. Barat: *Physics of Electronic and Atomic Collisions*, ed. by T. Andersen, B. Fastrup, F. Folkmann, H. Knudsen and N. Andersen (AIP, New York 1993) p. 537
[1.40] D.D. Briglia, D. Rapp: J. Chem. Phys. **42**, 3201 (1965);
D. Rapp, P. Englander-Golden: J. Chem. Phys. **43**, 1464 (1965)
[1.41] B.L. Schram, F.J. de Heer, M.J. van der Wiel, J. Kistemaker: Physica **31**, 94 (1965)
[1.42] K.D. Sevier: *Low Energy Electron Spectroscopy* (Wiley, New York 1972)
[1.43] N. Stolterfoht: Z. Phys. **248**, 81 (1971)
[1.44] N. Stolterfoht, D. Schneider, D. Burch, H. Wiesman, J.S. Risley: Phys. Rev. Lett. **33**, 59 (1974)
[1.45] S. Schumann, K.O. Groeneveld, G. Nolte, B. Fricke: Z. Phys. A **289**, 245 (1979)
[1.46] J.H. Posthmus, R. Morgenstern: J. Phys. B **25**, 4533 (1992)
[1.47] K. Wakiya: Unpublished (1997)
[1.48] R. Moshammer, M. Unverzagt, W. Schmitt, J. Ullrich, H. Schmidt-Böcking: Nucl. Instrum. Methods B **108**, 425 (1996)

[1.49] H. Kollmus, W. Schmitt, R. Moshammer, M. Unverzagt, J. Ullrich: Nucl. Instrum. Methods **124**, 377 (1997)
[1.50] A. Itoh, T. Schneider, G. Schwietz, Z. Roller, H. Platten, G. Nolte, D. Schneider, N. Stolterfoht: J. Phys. B **16**, 3965 (1983)
[1.51] N. Stolterfoht: Phys. Rep. **146**, 315 (1987)
[1.52] T.J.M. Zouros, D.H. Lee: Zero-degree Auger electron spectroscopy of projectile ions, in *Accelerator-Based Atomic Physics Techniques and Applications*, ed. by S. Shafroth and J.C. Austin (AIP, New York 1997)
[1.53] N. Stolterfoht, R.D. DuBois, R.D. Rivarola: Mechanisms for electron emission, in *Heavy Ion-Atom Collisions* (Springer, Berlin, Heidelberg 1977)
[1.54] H. Tawara, T. Iwai, Y. Kaneko, M. Kimura, N. Kobayashi, A. Matsumoto, S. Ohtani, K. Okuno, S. Takagi, S. Tsurubuchi: J. Phys. B **18**, 337 (1985)
[1.55] A. Remscheid, B.A. Huber, M. Pykavy, V. Schaemmler, K. Wiesemann: J. Phys. B **29**, 515 (1996)
[1.56] R. Schuch, S. Datz, P.F. Dittner, R. Hippler, H.F. Krause, P.D. Miller: Nucl. Instrum. Methods A **262**, 6 (1987)
R. Schuch, H. Schöne, P.D. Miller, H.F. Krause, D.F. Dittner, S. Datz, R.E. Olson: Phys. Rev. Lett. **60**, 925 (1988)
[1.57] W. Wu, S. Datz, N.L. Jones, H.F. Krause, B. Rosner, K.D. Sorge, C.R. Vane: Phys. Rev. Lett. **76**, 4234 (1996)
[1.58] I. Katayama, H. Ikegami, H. Ogawa, Y. Haruyama, M. Tozaki, A. Aoki, F. Fukusawa, K. Yoshida, I. Sugai: Phys. Rev. A **53**, 242 (1996)
[1.59] M. Barat, M.N. Gaboriaud, L. Guillemont, P. Roncin, H. Laurent, S. Andriamonje: J. Phys. B **20**, 5771 (1987)
[1.60] J.T. Park, F.D. Schowengerdt: Rev. Sci. Instr. **40**, 753 (1969); Phys. Rev. **185**, 152 (1969)
[1.61] W.C. Wiley, I.H. McLaren: Rev. Sci. Instr. **26**, 1150 (1955)
[1.62] J. Ullrich, H. Schmidt-Böcking: Phys. Lett. A **125**, 193 (1987)
[1.63] M.F.A. Harrison: Brit. J. Appl. Phys. **17**, 371 (1966)
[1.64] D.F. Dance, M.F.A. Harrison, A.C.H. Smith: Proc. Roy. Soc. A **290**, 74 (1966)
[1.65] D.H. Crandall, R.A. Phaneuf, P.O. Taylor: Phys. Rev. A **18**, 1911 (1978)
[1.66] D.W. Hughes, R.K. Feeney: Phys. Rev. A **23**, 2241 (1981)
[1.67] P. Defrance, F. Brouillard, W. Claeys, G. Van Wassenhove: J. Phys. B **14**, 103 (1981)
[1.68] P. Defrance, W. Claeys, A. Cornet, G. Poulaert: J. Phys. B **14**, 111 (1981)
[1.69] H.B. Gilbody, R. Browning, G. Levy, A.L. McIntosh, K.F. Dunn: J. Phys. B **1**, 863 (1968)
[1.70] H. Tawara: J. Phys. Soc. Jpn. **31**, 871 (1971)
[1.71] E. Brook, M.F.A. Harrison, A.C. Smith: J. Phys. B **11**, 3115 (1978)
[1.72] A.J. Dixon, M.F.A. Harrison, A.C.H. Smith: J. Phys. B **9**, 2617 (1976)
[1.73] M.E. Lagus, J.B. Boffard, L.W. Anderson, C.C. Lin: Phys. Rev. A **53**, 1505 (1996)
[1.74] J.B. Mitchell, K.F. Dunn, G.C. Angel, R. Browning, H.B. Gilbody: J. Phys. B **10**, 1897 (1977)
[1.75] K. Rinn, F. Melchert, E. Salzborn: J. Phys. B **18**, 3783 (1985)
[1.76] J.H. Posthmus, R. Morgenstern: Phys. Rev. Lett. **68**, 1315 (1992) ; J. Phys. B **25**, 4533 (1992)
[1.77] G. de Nijs, R. Hoekstra, R. Morgenstern: J. Phys. B **27**, 2557 (1994)
[1.78] J.A. Tanis, S.M. Shafroth, J.E. Willis, M. Clark, J.K. Swenson, E.N. Strait, J.R. Mowat: Phys. Rev. Lett. **47**, 828 (1981)

[1.79] M. Clark, D. Brandt, J.K. Swenson, S.M. Shafroth: Phys. Rev. Lett. **54**, 454 (1985)
[1.80] D. Brandt: Phys. Rev. A **27**, 1314 (1983)
[1.81] J.K. Swenson, Y. Yamazaki, P.D. Miller, H.F. Krause, P.F. Dittner, P.L. Pepmiller, S. Datz, N. Stolterfoht: Phys. Rev. Lett. **57**, 3042 (1986)
[1.82] M. Schulz, E. Justiniano, R. Schuch, P.H. Mokler, S. Reusch: Phys. Rev. Lett. **58**, 1734 (1987)
[1.83] P. Pepmiller, P. Richard, J. Newcomb, J. Hall, T.R. Dillingham: Phys. Rev. A **31**, 734 (1985)
[1.84] R. Dörner, J. Ullrich, O. Jaguzki, S. Lencinas, A. Gensmantel, H. Schmidt-Böcking: In *Electronic and Atomic Collisions*, ed. by W.R. MacGillivary, I.E. McCarthy and M.C. Standage (Hilger, Bristol 1992) p. 351
[1.85] R. Ali, V. Fröhne, C.L. Cocke, M. Stöckli, S. Cheng, M.L.A. Raphaelian: Phys. Rev. Lett. **69**, 2491 (1992)
[1.86] W. Wu, J.P. Giese, Z. Chen, R. Ali, C.L. Cocke, P. Richard, M. Stöckli: Phys. Rev. A **50**, 502 (1994)
[1.87] L.A. Raphaelian, M. Stöckli, W. Wu, C.L. Cocke: Phys. Rev. A **51**, 1304 (1995)
[1.88] V. Mergel, R. Dörner, J. Ullrich, O. Jagutzki, S. Nüttgens, L. Spielberger, M. Unverzagt, C.L. Cocke, R.E. Olson, M. Schulz, U. Buck, E. Zanger, W. Theisinger, M. Isser, S. Geis, H. Schmidt-Böcking: Phys. Rev. Lett. **74**, 2200 (1995)
[1.89] R. Moshammer, J. Ullrich, H. Kollmus, W. Schmitt, M. Unverzagt, O. Jagutzki, V. Mergel, H. Schmidt-Böcking, R. Mann, C.J. Woods, R.E. Olson: Phys. Rev. Lett. **77**, 1242 (1996)
[1.90] T. Kambara, J.Z. Tang, Y. Awaya, B.D. Depaola, O. Jagutzki, Y. Kanai, M. Kimura, T.M. Kojima, V. Mergel, Y. Nakai, H. Schmidt-Böcking, I. Shimamura: J. Phys. B **28**, 4593 (1995)
[1.91] J. Ullrich, R. Moshammer, R. Dörner, O. Jagutzki, V. Mergel, H. Schmidt-Böcking, L. Spielberger: J. Phys. B **29**, 2917 (1997)
[1.92] S.D. Kravis, M. Abdallah, C.L. Cocke, C.D. Lin, M.Stöckli, B. Walch, Y.D. Wang, R.E. Olson, V.D. Rodriguez, W. Wu, M. Pieksma, N. Watanabe: Phys. Rev. A **54**, 1394 (1996)

Chapter 2

[2.1] L.J. Kieffer, G.H. Dunn: Rev. Mod. Phys. **38**, 135 (1966)
[2.2] T.D. Märk, G.H. Dunn (eds.): *Electron Impact Ionization* (Springer, Berlin, Heidelberg 1985)
[2.3] F. Brouillard (ed): *Atomic Processes in Electron-Ion and Ion-Ion Collisions*, NATO ASI Series B, No.145 (Plenum, New York 1986)
[2.4] A. Müller: In *Physics of Ion Impact Phenomena*, ed. D. Mathur, Springer Ser. Chem. Phys. Vol. 54 (Springer, Berlin, Heidelberg 1991)
[2.5] J.H. McGuire: Adv. At. Mol. Opt. Phys. **29**, 217 (1992)
[2.6] D.L. Moores, K.J. Reed: Adv. At. Mol. Opt. Phys. **34**, 301 (1994)
[2.7] P. Defrance, M. Duponchelle, D.L. Moores: In *Atomic and Molecular Processes in Fusion Edge Plasma* ed. by R.K. Janev (Plenum, New York 1995)
[2.8] J.H. McGuire: In *Atomic Inner Shell Processes* ed. by B. Crasemann (Academic, New York 1995) Chap. 7

[2.9] J.H. McGuire, J.C. Straton, T. Ishihara: In *Atomic, Molecular and Optical Physics Reference Book* ed. by G.W.F. Drake (AIP, New York 1996) Chap. 40
[2.10] J.H. McGuire: *Introduction to Dynamic Correlation: Multiple Electron Transitions in Atomic Collisions* (Tulane University 1997)
[2.11] H. Tawara, T. Kato: At. Data Nucl. Data Tables **36**, 167 (1987)
[2.12] V.P. Shevelko, H. Tawara, E. Salzborn: *Mutiple-Ionization Cross Sections of Atoms and Positive Ions by Electron Impact*, Report NIFS-DATA-27, (National Institute for Fusion Science, Nagoya, Japan 1995)
[2.13] V.P. Shevelko, H. Tawara: J. Phys. B **28**, L589 (1995)
[2.14] C. Bélenger, P. Defrance, E. Salzborn, V.P. Shevelko, H. Tawara, D.B. Uskov: J. Phys. B **30**, 2667 (1997)
[2.15] W. Lotz: J. Opt. Soc. Am. **59**, 915 (1968); *ibid*, **60**, 206 (1970)
[2.16] T.A. Carlson, C.W. Nestor, Jr., N. Wasserman, J.D. McDowell: At. Data **2**, 63 (1970)
[2.17] M.A. Bolorizadeh, C.J. Patton, M.B. Shah, H.B. Gilbody: J. Phys. B **27**, 175 (1994)
[2.18] R.S. Freund, R.C. Wetzel, R.J. Shul, T.R. Hayes: Phys. Rev. A **41**, 3575 (1990)
[2.19] W. Lotz: Z. Phys. **232**, 101 (1970)
[2.20] R.E. Fox: J. Chem. Phys. **33**, 200 (1960)
[2.21] K. Dolder: Adv. At. Mol. Opt. Phys. **32**, 69 (1994)
[2.22] E. Krishnakumar, S.K. Srivastava: J. Phys. B **21**, 105 (1988)
[2.23] D. Almeida, A.C. Fontes, C.F.L. Godinho: J. Phys. B **28**, 3335 (1995)
[2.24] P. McCallion, M.B. Shah, H.B. Gilbody: J. Phys. B **25**, 1051 (1992)
[2.25] L.A. Vainshtein, V.I. Ochkur, V.I. Rakhovskii, A.M. Stepanov: Sov. Phys. - JETP, **34**, 271 (1972)
[2.26] S. Okudaira, Y. Kaneko, I. Kanomata: J. Phys. Soc. Jpn. **28**, 1536 (1970)
[2.27] F. Karstensen, M. Schneider: J. Phys. B **11**, 167 (1978)
[2.28] E.M. Oualim: "Etude Experimentale de L'Ionization des Ions Kr^{q+} (q = 7 á 13) par Impact d'Electrons"; PhD Thesis, Université Catholique de Louvain, Belgium (1995) (unpublished)
[2.29] E.M. Oualim, M. Duponchelle, P. Defrance: Nucl. Instrum. Methods B **98**, 150 (1995)
[2.30] M. Zambra, D. Belic, P. Defrance: J. Phys. B **27**, 2383 (1994)
[2.31] V. Fisher, Yu. Ralchenko, A. Goldrich, D. Fisher, Y. Maron: J. Phys. B **28**, 3027 (1995)
[2.32] H. Deutsch, K. Becker, T.D. Märk: Control. Plama Phys. **35**, 421 (1995); J. Phys. B **29**, L497 (1996)
[2.33] C.J. Patton, K.O. Lozhkin, M.B. Shah, J. Geddes, H.B. Gilbody: J. Phys. B **29**, 1409 (1996)
[2.34] B.L. Schram: Physica **32**, 197 (1966)
[2.35] H. Lebius, J. Binder, H.R. Kozlowski, K. Wiesemann, B.A. Huber: J. Phys. B **22**, 83 (1989)
[2.36] J.A. Syage: Phys. Rev. A **46**, 5666 (1992)
[2.37] M.K. Stenke, D. Hathiramani, G. Hofmann, V.P. Shevelko, M. Steidl, R. Völpel, E. Salzborn: Nucl. Instrum. Methods B **98**, 138 (1995)
[2.38] A.M. Howald, D.C. Gregory, F.W. Meyer, R.A. Phaneuf, A. Müller, N. Djuric, G.H. Dunn: Phys. Rev. A **33**, 3779 (1986)
[2.39] A. Müller, R. Frodl: Phys. Rev. Lett. **44**, 29 (1980)

[2.40] A. Müller, C. Achenbach, E. Salzborn, R. Becker: J. Phys. B **17**, 1427 (1984)
[2.41] K. Tinschert, A. Müller, R. Becker, E. Salzborn: J. Phys. B **20**, 1823 (1987)
[2.42] K. Tinschert, A. Müller, R.A. Phaneuf, G. Hofmann, E. Salzborn: J. Phys. B **22**, 1241 (1989)
[2.43] V.P. Shevelko, H. Tawara: In *Atomic and Plasma-Material Interaction Data for Fusion* (Nucl. Fusion Suppl., ed. by R.K. Janev) Vol. **6**, 101 (1995)
[2.44] L.H. Andersen, P. Hvelplund, H. Knudsen, S.P. Møller, A.H. Sørensen: Phys. Rev. A **36**, 3612 (1987)
[2.45] B. El Marji, A. Lahmam-Bennani, A. Duguet, T.J. Reddish: J. Phys. B **29**, L 157 (1996)
[2.46] R. Mkhanter, C. Dal Cappollo, A. Lahmam-Bennani, Yu.V. Popov: J. Phys. B **29**, 1101 (1996)
[2.47] P. Lamy, B. Joulakian, C. Dal Cappollo, A. Lahmam-Bennani: J. Phys. B **29**, 2315 (1996)
[2.48] J. Berakdar: Phys. Lett. A **220**, 237 (1996))
[2.49] M. Gryzinski: Phys. Rev. A **138**, 336 (1965)
[2.50] F.N. Byron, C.J. Joachain: Phys. Rev. Lett. **16**, 1139 (1966); Phys. Rev. A **164**, 1 (1967)
[2.51] R.J. Tweed: J. Phys. B **5**, 256 (1973); *ibid*, **6**, 270 (1973)
[2.52] P. Grujic: J. Phys. B **16**, 2567 (1983)
[2.53] B.L. Schram, A.J.H. Boerboom, J. Kistemaker: Physica **32**, 185 (1966)
[2.54] T.D. Märk: Beitr. Plasmaphys. **22**, 257 (1982)
[2.55] R.C. Wetzel, F.A. Baiocchi, T.R. Hayes, R.S. Freund: Phys. Rev. A **35**, 559 (1987)
[2.56] M. Steidl, D. Hathiramani, G. Hofmann, M. Stenke, R. Völpel, E. Salzborn: XIX ICPEAC, Whistler, BC (1995) Book of Abstracts, p. 564
[2.57] J.H. McGuire: Phys. Rev. Lett. **49**, 1153 (1982)
[2.58] J.D. Jackson: *Classical Electrodynamics*, 2nd ed. (Wiley, New York 1975)
[2.59] H.A. Bethe, E.E. Salpeter: *Quantum Mechanics of One- and Two-Electron Atoms*, (Plenum, New York 1977)
[2.60] M.A. Kornberg, J.E. Miraglia: Phys. Rev. A **49**, 5120 (1994)
[2.61] D.B. Uskov: Invited Papers of XIX ICPEAC, **360**, 687 (AIP, New York 1995)
[2.62] A.L. Ford, J.F. Reading: J. Phys. B **23**, 3131 (1990)
[2.63] M.B. Shah, D.S. Elliott, P. McCallion, H.B. Gilbody: J. Phys. B **21**, 2751 (1988)
[2.64] D.J. Yu, S. Rachafi, J. Jureta, P. Defrance: J. Phys. B **25**, 4593 (1992)
[2.65] B. Peart, K.T. Dolder: J. Phys. B **2**, 1169 (1969)
[2.66] A. Müller, W. Groh, U. Kneissl, R. Heil, H. Ströher, E. Salzborn: J. Phys. B **16**, 2039 (1983)
[2.67] D.H.H. Hoffmann, C. Brendel, H. Genz, W. Low, S. Müller, A. Richter: Z. Phys. A **293**, 187 (1979)
[2.68] T.A. Carlson, W.E. Hunt, M.O. Krause: Phys. Rev. **151**, 41 (1966)
[2.69] R. Mann: Z. Phys. D **3**, 85 (1985)
[2.70] S. Kelbch, J. Ullrich, R. Mann, P. Richard, H. Schmidt-Böcking: J. Phys. B **18**, 323 (1985)
[2.71] H. Schmidt-Böcking, U. Ramm, G. Kraft, J. Ullrich, H. Berg, C. Kelbch, R.E. Olson, R. DuBois, S. Hagmann, F. Jiazheni: Adv. Space Res. **12**, 2 (1992)

[2.72] K.-G. Dietrich, D.H. Hoffmann, E. Boggasch, J. Jacoby, H. Wahl, M. Elfers, C.R. Haas, V.P. Dubenkov, A.A. Golubev: Phys. Rev. Lett. **69**, 3623 (1992)
[2.73] P.V. Dressendorfer: Nucl. Instrum. Methods B **40/41**, 1291 (1989)
[2.74] M. Yoshida, M. Tamada, M. Asano, H. Omichi, H. Kubota, R. Katakai, R. Spohr, J. Vetter: Rad. Eff. Def. Solids **126**, 409 (1993)
[2.75] T. Matsuo, T. Tonuma, H. Kumagai, H. Tawara: Phys. Rev. A **50**, 1178 (1994)
[2.76] J.H. McGuire: J. Phys. B **28**, 913 (1995)
[2.77] J.F. Reading, A.L. Ford: Phys. Rev. Lett. **58**, 543 (1987); J. Phys. B **20**, 3747 (1987)
[2.78] A.L. Ford, L.A. Wehrmann, K.A. Hall, J.F. Reading: J. Phys. B **30**, 2889 (1997)
[2.79] L.P. Presnyakov, H. Tawara, I.Yu. Tolstikhina, D.B. Uskov: J. Phys. B **28**, 785 (1995)
[2.80] L.A. Presnyakov, D.B. Uskov: Sov. Phys. - JETP Lett. **66**, 7 (1997)
[2.81] D.B. Uskov: Phys. Scr. T **73**, 133 (1997)
[2.82] J. Ullrich, R. Dörner, H. Berg, C.L. Cocke, J. Euler, K. Froschauer, S. Hagmann, O. Jagutzki, S. Lencinas, R. Mann, V. Mergel, R. Moshammer, H. Schmidt-Böcking, H. Tawara, M. Unverzagt: Nucl. Instrum. Methods B **87**, 70 (1994)
[2.83] H. Knudsen, L.H. Andersen, P. Hvelplund, G. Astner, H. Cederquist, H. Danared, L. Liljeby, K.G. Rensfelt: J. Phys. B **17**, 3545 (1984)
[2.84] M.B. Shah, H.B. Gilbody: J. Phys. B **18**, 899 (1984)
[2.85] E. Krishnakumar, F.A. Radjgara: J. Phys. B **26**, 4155 (1993)
[2.86] C.L. Cocke, R.E. Olson: Phys. Rep. **205**, 153 (1991)
[2.87] T.J. Gray, C.L. Cocke, E. Justiniano: Phys. Rev. A **22**, 849 (1980)
[2.88] S. Kelbch, H. Schmidt-Böcking, J. Ullrich, R. Schuch, E. Justiniano, M. Ingwersen, C.L. Cocke: Z. Phys. A **317**, 9 (1984)
[2.89] J. Ullrich, K. Bethge, S. Kelbch, W. Schadt, H. Schmidt-Böcking, K.E. Stiebing: J. Phys. B **19**, 448 (1986)
[2.90] A. Müller, B. Schuch, W. Groh, E. Salzborn, H.F. Beyer, P.M. Mokler, R.E. Olson: Phys. Rev. A **33**, 3010 (1986)
[2.91] A. Müller, B. Schuch, W. Groh, E. Salzborn: Z. Phys. D **7**, 251 (1987)
[2.92] S. Kelbch, C.L. Cocke, S. Hagmann, M. Horbatsch, C. Kelbch, R. Koch, H. Schmidt-Böcking, J. Ullrich: J. Phys. B **23**, 1277 (1989)
[2.93] D.R. Schulz, R.E. Olson, C.O. Reinhold, S. Kelbch, H. Schmidt-Böcking, J. Ullrich: J. Phys. B **23**, 3839 (1990)
[2.94] T. Tonuma, H Kumagai, T. Matsuo, H. Tawara: Phys. Rev. A **40**, 6328 (1989)
[2.95] H. Tawara, T. Tonuma, H. Kumagai, T. Matsuo: Phys. Rev. A **41**, 116 (1990)
[2.96] G. de Nijs, H.O. Folkerts, R. Hoekstra, R. Morgenstern: J. Phys. B **29**, 85 (1996)
[2.97] S. Wexler: J. Chem. Phys. **41**, 1714 (1964)
[2.98] A.S. Schlachter, K.H. Berkner, H.F. Beyer, W.G. Graham, W. Groh, R. Mann, A. Müller, R.E. Olson, R.V. Pyle, J.W. Stearns, J.A. Tanis: Phys. Scr. **3**, 153 (1983)

[2.99] S. Kelbch, J. Ullrich, W. Rauch, H. Schmidt-Böcking, M. Horbatsch, R.M. Dreizler, S. Hagmann, R. Anholt, A.S. Schlachter, A. Müller, P. Richard, Ch. Stoller, C.L. Cocke, R. Mann, W.E. Meyerhof, J.D. Rasmussen: J. Phys. B **19**, L47 (1986)
[2.100] R.D. DuBois, S.T. Manson: Phys. Rev. A **35**, 2007 (1987)
[2.101] R.D. DuBois: Phys. Rev. A **36**, 2585 (1987)
[2.102] R.D. DuBois: Phys. Rev. A **39**, 4440 (1989)
[2.103] R.E. Olson, J. Ullrich, H. Schmidt-Böcking: J. Phys. B **20**, L809 (1987)
[2.104] C.L. Cocke: Phys. Rev. A **20**, 749 (1979)
[2.105] S. Kelbch, C.L. Cocke, S. Hagmann, M. Horbatsch, C. Kelbch, R. Koch, H. Schmidt-Böcking, J. Ullrich: J. Phys. B **23**, 1277 (1990)
[2.106] H. Berg: Report GSI-93-12 (Darmstadt, Germany 1993)
[2.107] J. Ullrich, C.L. Cocke, S. Kelbch, R. Mann, P. Richard, H. Schmidt-Böcking: J. Phys. B **17**, L785 (1984)
[2.108] M.B. Shah, C.J. Patton, M.A. Bolorizadeh, H.B. Gilbody: J. Phys. B **28**, 1821 (1995)
[2.109] C.J. Patton, M.B. Shah, M.A. Bolorizadeh, J. Geddes, H.B. Gilbody: J. Phys. B **28**, 3889 (1995)
[2.110] M.B. Shah, C.J. Patton, J. Geddes, H.B. Gilbody: Nucl. Instrum. Methods B **98**, 280 (1995)
[2.111] O. Heber, G. Sampoll, B.B. Bandong, R.J. Maurer, R.L. Watson, I. Ben-Itzhak, J.M. Sanders, J.L. Shinpaugh, P. Richard: Phys. Rev. A **52**, 4578 (1995)
[2.112] K.O. Lozhkin, C.J. Patton, P. McCartney, M. Santánna, M.B. Shah, J. Geddes, H.B. Gilbody: J. Phys. B **30**, 1785 (1997)
[2.113] E.W. McDaniel, J.B.A. Mitchell, M.E. Rudd: In *Atomic Collisions* (Heavy Particle Projectiles) (Wiley, New York 1993)
[2.114] M. Horbatsch: Z. Phys. D **21**, S 63 (1991)
[2.115] P. Curruthers, F. Zachariasen: Rev. Mod. Phys. **55**, 245 (1983)
[2.116] I. Ben-Itzhak, T.G. Gray, J.C. Legg, J.H. McGuire: Phys. Rev. A **37**, 3685 (1988)
[2.117] R.E. Olson, J. Ullrich, R. Dörner, H. Schmidt-Böcking: Phys. Rev. A **40**, 2843 (1989)
[2.118] J.H. McGuire, L. Weaver: Phys. Rev. A **16**, 41 (1977)
[2.119] O. Heber, R.L. Watson, G. Sampoll, B.B. Bandong: Phys. Rev. A **42**, 6466 (1990)
[2.120] V.A. Sidorovich: J. Phys. B **14**, 4805 (1981)
[2.121] V.A. Sidorovich, V.S. Nikolaev, J.H. McGuire: Phys. Rev. A **31**, 2193 (1985)
[2.122] Yu.S. Sayasov: J. Phys. B **26**, 1197 (1993)
[2.123] A. El-Shemi, Y. Loftes, G. Zschornack: J. Phys. B **30**, 237 (1997)
[2.124] V. Krishnamurthi, I. Ben-Itzhak, K.D. Carnes: J. Phys. B **29**, 287 (1996)
[2.125] I. Lesteven-Vaisse, D. Hennecart, R. Gayet: J. Phys. (France), **49**, 1529 (1988)
[2.126] R. Gayet: J. Phys. (France), **50**, C1-53 (1989)
[2.127] A. Salin: J. Phys. B **22**, 3901 (1989)
[2.128] R.E. Olson: in: *Electronic and Atomic Collisions* (Elsevier, Amsterdam 1988)
[2.129] R.E. Olson, A. Salop: Phys. Rev. A **16**, 531 (1977)
[2.130] R.E. Olson: Phys. Rev. A **27**, 1871 (1983)

[2.131] L. Nagy, J.H. McGuire, L. Végh, B. Sulik, N. Stolterfoht: J. Phys. B **30**, 1939 (1997)
[2.132] A. Russek, M.T. Thomas: Phys. Rev. **114**, 1538 (1959)
[2.133] A. Russek: Phys. Rev. **132**, 246 (1963)
[2.134] A. Russek, J. Meli: Physica **46**, 222 (1970)
[2.135] H. Berg, R. Dörner, C. Kelbch, S. Kelbch, J. Ullrich, S. Hagmann, P. Richard, H. Schmidt-Böcking, A.S. Schlachter, M. Prior, H.J. Crawford, J.M. Engelage, I. Flores, D.H. Lyod, J. Pedersen, R.E. Olson: J. Phys. B **21**, 3929 (1988)
[2.136] A. Müller, W. Groh, E. Salzborn: Phys. Rev. Lett. **51**, 107 (1983)
[2.137] M. Horbatsch, R.M. Dreizler: Z. Phys. D **2**, 183 (1986);
M. Horbatsch: Z. Phys. D **1**, 611 (1986); M. Horbatsch: J. Phys. B **19**, L193 (1986)
[2.138] J. Ullrich, M. Horbatsch, V. Dangendorf, S. Kelbch, H. Schmidt-Böcking: J. Phys. B **21**, 611 (1988)
[2.139] M. Horbatsch: J. Phys. B **25**, 3797 (1992)
[2.140] M. Horbatsch. R.M. Dreizler: Phys. Lett. A **113**, 251 (1985)
[2.141] V.P. Shevelko, D.B. Uskov, H. Tawara: Book of Abstr. of XX ICPEAC, Vienna 1997, p. WE164 (Tech. Univ., Vienna 1997)
[2.142] F. Martín, A. Salin: Phys. Rev. Lett. **76**, 1437 (1996)
[2.143] R. Bruch, P.L. Altick, E. Träbert, P.H. Heckmann: J. Phys. B **17**, L655 (1984);
R. Bruch, L. Kochach, E. Träbert, P.H. Heckmann, B. Raith, U. Will: Nucl. Instrum. Methods B **9**, 438 (1985)
[2.144] J.O.P. Pedersen, P. Hvelplund: Phys. Rev. Lett. **62**, 2373 (1989)
[2.145] J.P. Giese, M. Schulz, J.K. Swenson, H. Schöne, M. Benhenni, S.L. Varghese, C.R. Vane, P.F. Dittner, S.M. Shafroth, S. Datz: Phys. Rev. A **42**, 1231 (1990)
[2.146] A.V. Vinogradov: *Quasiclassical method in collision theory of atoms with electrons and heavy particles*, Proc. P.N. Lebedev Physics Institute, Vol. **51**, p. 44 (Moscow 1970)
[2.147] I.L. Beigman: *Effective cross sections calculated with account for configuration interaction*, Proc. P.N. Lebedev Physics Institute, Vol. **51**, p. 55 (Moscow 1970)
[2.148] H. Bachau, M. Bahri, F. Martin, A. Salin: J. Phys. B **24**, 2015 (1991)
[2.149] J.C. Stratton, J.H. McGuire, Z. Chen: Phys. Rev. A **46**, 5514 (1992)
[2.150] A.L. Godunov, N.V. Novikov, V.S. Senashenko: J. Phys. B **25**, L43 (1992);
A.L. Godunov, V.A. Shipakov: J. Phys. B **26**, L811 (1993)
[2.151] W. Fritsch, C.D. Lin: Phys. Rev. A **41**, 4776 (1990)
[2.152] J.H. McGuire, J.C. Straton: Phys. Rev. A **43**, 5184 (1990)
[2.153] T.G. Winter: Phys. Rev. A **43**, 4727 (1991)
[2.154] K. Moribayashi, K. Hino, M. Matzusawa, M. Kimura: Phys. Rev. A **44**, 7234 (1991); *ibid*, **45**, 7922 (1992); *ibid*, **46**, 1684 (1992); *ibid*, **47**, 4874 (1993)
[2.155] H.A. Slim, B.H. Bransden, D.R. Flower: J. Phys. B **26**, L159 (1993)
[2.156] V.S. Sidorovich: Phys. Scr. **50**, 119 (1994)
[2.157] F.W. Byron, C.J. Joachain: Phys. Rev. Lett. **16**, 1139 (1966)
[2.158] A. Chatioui: Unpublished (1990)
[2.159] K. Hino, H. Okamoto, M. Matsuzawa, M. Kimura: Phys. Rev. A **49**, 3753 (1994)
[2.160] J. van Eck: *Thesis*, University of Gemeente, The Netherlands (1964)

[2.161] R. Bruch, D. Schneider, W.H.E. Schwartz, M. Meinhart, B.M. Johnson, K. Taubjerg: Phys. Rev. A **19**, 587 (1979)
[2.162] R. Bruch, S. Fülling: unpublished (1990)
[2.163] J.O.P. Pedersen, F. Folkmann: J. Phys. B **23**, 441 (1990)
[2.164] S. Fülling, R. Bruch, E.A. Rauscher, P.A. Neil, E. Träbert, P.H. Heckmann, J.H. McGuire: Nucl. Instru. Methods **56/57**, 275 (1972); Phys. Rev. Lett. **68**, 3152 (1992)
[2.165] M. Bailey, R. Bruch, E. Rauscher, S. Bliman: J. Phys. B **28**, 2655 (1995)
[2.166] A.K. Edwards, R.M. Wood, J.L. Davis, R.L. Ezell: Phys. Rev. A **42**, 1367 (1990);
A.K. Edwards, R.M. Wood, J.L. Davis, R.L. Ezell: Phys. Rev. A **44**, 797 (1991);
A.K. Edwards, R.M. Wood, M.A. Mangan, R.L. Ezell: Phys. Rev. A **46**, 6970 (1992)
[2.167] J.L. Forand: J. Phys. B **18**, 1409 (1985)
[2.168] E. Träbert, P.H. Heckmann, R. Bruch, S. Fülling: Nucl. Instrum. Methods B **23**, 151 (1987)
[2.169] L.A. Vainshtein, A.V. Vinogradov: Opt. Spectrosc. **23**, 185 (1967)
[2.170] I.R. Taylor, K.L. Bell, A.E. Kingston: J. Phys. B **13**, 2983 (1980)
[2.171] R. Bruch, L. Kocbach, E. Träbert, P.H. Heckmann, B. Raith, U. Will: Nucl. Instrum. Methods B **9**, 438 (1985)
[2.172] R. Bruch, I.L. Beigman, E.A. Rauscher, S. Fülling, J.H. McGuire, E.Träbert, P.H. Heckmann: J. Phys. B **26**, L413 (1993)
[2.173] L. Nagy, J. Wang, J.C. Straton, J.H. McGuire: Phys. Rev. A **52**, R902 (1995)
[2.174] V.S. Sidorovich: J Phys. B **30**, 2187 (1997)
[2.175] M.R.H. Rudge: J. Phys. B **21**, 1887 (1988)
[2.176] A. Raeker, K. Bartschaft, R.H.G. Reid: J. Phys. B **27**, 3129 (1994)
[2.177] K.H. Schartner, B. Lommel, D. Dettleffsen: J. Phys. B **24**, L13 (1991)
[2.178] B. Lommel: *Thesis*, Giessen University, Germany (1992)
[2.179] H. Klinger, A. Müller, E. Salzborn: J. Phys. B **8**, 230 (1975)
[2.180] E. Salzborn: IEEE Trans. NS-**23**, 947 (1976)
[2.181] A. Müller, E. Salzborn: Phy. Lett. A **62**, 391 (1977)
[2.182] B.R. Beck, J. Steiger, G. Weienberg, D.A. Church, J. McDonald, D. Schneider: Phys. Rev. Lett. **77**, 1735 (1996)
[2.183] M. Kimura, H. Nakamura, H. Watanabe, I. Yamada, A. Danjo, K. Kosaka, A. Matsumoto, S. Chtani, H.A. Sakane, M. Sakurai, H. Tawara, M. Yoshino: J. Phys. B **28**, L643 (1995)
[2.184] H. Winter, Th.M. El-Sherbini, E. Bloemen, F.J. de Heer, A. Salop: Phys. Lett. A **68**, 211 (1978)
[2.185] Th.M. El-Sherbini, A. Salop, E. Bloemen, F.J. DeHeer: J. Phys. B **12**, L579 (1979)
[2.186] C.L. Cocke, R.D. DuBois, T.J. Gray, E. Justiniano, C. Can: Phys. Rev. Lett. **46**, 1671 (1981)
[2.187] E. Justiniano, C.L. Cocke, T.J. Gray, R.D. DuBois, C. Can: Phys. Rev. A **24**, 2953 (1981)
[2.188] E.H. Nielsen, L.H. Andersen, A. Bárány, H. Cederquist, P. Hvelplund, H. Knudsen, K. MacAdam, J. Sorensen: J. Phys. B **17**, L139 (1984);
E.H. Nielsen, L.H. Andersen, A. Bárány, H. Cederquist, J. Heinemeier, P. Hvelplund, H. Knudsen, K. MacAdam, J. Sorensen: J. Phys. B **18**, 1789 (1985)

[2.189] S. Kravis, H. Saitoh, K. Okuno, K. Soejima, M. Kimura, I. Shimamura, Y. Awaya, Y. Kaneko, M. Oura, N. Shimakura: Phys. Rev. A **52**, 1206 (1995)

[2.190] K. Okuno, H. Saitoh, S. Kravis, N. Kobayashi: AIP Conf. Proc. **360**, 867 (AIP, New York 1995)

[2.191] K. Okuno, H. Saitoh, K. Soejima, S.D. Kravis, N. Kobayashi: Phys. Scr. **71**, 140 (1997)

[2.192] K. Suzuki, K. Okuno, N. Kobayashi: Phys. Scr. **73**, 172 (1997)

[2.193] R.E. Olson, A. Salop: Phys. Rev. A **14**, 579 (1976)

[2.194] G. Gioumousis, D.P. Stevenson: J. Chem. Phys. **29**, 294 (1958)

[2.195] K. Sato, K. Takiyama, T. Oda, U. Furukane, R. Akiyama, M. Miura, M. Otsuka, H. Tawara: J. Phys. B **27**, L652 (1994)

[2.196] Ya.M. Fogel, R.V. Mitin, V.F. Kozlov, N.D. Romashko: Sov. Phys. - JETP **8**, 390 (1959)

[2.197] V.S. Nikolaev, I.S. Dmitriev, L.N. Fateeva, Ya. A. Teplova: Sov. Phys. - JETP **12**, 627 (1961)

[2.198] U. Schryber: Helv. Phys. Acta **40**, 1023 (1967)

[2.199] T.C. Thiesen, J.H. McGuire: Phys. Rev A **20**, 1406 (1979)

[2.200] L.H. Toburen, M.Y. Nakai, R.A. Langly: Oak Ridge Natl. Lab. Report No. ORNL-TM-1988 (Oak Ridge, 1968)

[2.201] F. Fremont, H. Merabet, J.Y. Chesnel, X. Husson, A. Lepoutre, D. Lecler, G. Rieger, N. Stolterfoht: Phys. Rev A **50**, 3117 (1994)

[2.202] J.Y. Chesnel, H. Merabet, F. Fremont, G. Cremer, X. Husson, D. Lecler, G. Rieger, S. Spieler, M. Grether, N. Stolterfoht: Phys. Rev. A **53**, 4198 (1996)

[2.203] S. Martin, J. Bernard, L. Chen, A. Denis, J. Désesquelles: Phys. Scr. **73**, 149 (1997)

[2.204] J.-Y. Chesnel, M. Grether, A. Spieler, F. Frémont, C. Bedouet, X. Husson, H. Merabet, N. Stolterfoht: XX ICPEAC, Vienna (1997)

[2.205] J. Bernard, A. Chen, J. Désesquelles, S. Martin: Phys. Scr. **56**, 26 (1997)

[2.206] M. Barat, P.J. Roncin: J. Phys. B **25**, 2205 (1992)

[2.207] M. Barat, M.N. Gaboriaud, P.J. Roncin: Phys. Scr. T **46**, 210 (1993)

[2.208] A. Delon, S. Martin, A. Denis, Y. Querdance, M. Carre, J. Desesquelle, M.C. Buchet-Pouliazac: Rad. Eff. Def. in solids, **126**, 337 (1993)

[2.209] N. Bohr, J.K. Lindhard: Dan. Vidensk. Selsk. Mat.-Phys. Medd. **28/7**, (1954)

[2.210] A. Bárány, G. Astner, H. Cederquist, D. Danared, S. Holt, P. Hvelplund, A. Johnson, H. Knudsen, L. Liljeby, K.G. Rensfelt: Nucl. Instrum. Methods B **9**, 397 (1985)

[2.211] A. Niehaus: J. Phys. B **19**, 2925 (1986)

[2.212] C. Harel, H. Jouin: Europhys. Lett. **11**,21 (1990); J. Phys. B **25**, 2267 (1992)

[2.213] Z. Chen, R. Shingal, C.D. Lin: J. Phys. B **24**, 4215 (1991)

[2.214] J.P. Hansen, K. Taulbjerg: Phys. Rev. A **45**, R4214 (1992)

[2.215] V.K. Nikulin, N.A. Guschina: Phys. Scr. T **71**, 134 (1997)

[2.216] P. Moretto-Capelle, P. Benoit-Cattin, A. Bordenave-Montesquieu, A. Gleizes: Z. Phys. D., Suppl. **21**, S283 (1991)

[2.217] R. Mann: Multiple electron capture, in *Proc. Niels Bohr Centennial Conference*, Copenhagen (1985), ed. by J. Bang and J. De Boer (North Holland, Amsterdam 1985)

[2.218] J.R. Macdonald, F.W. Martin: Phys. Rev. A **4**, 1965 (1971); *ibid*, **4**, 1974 (1971)
[2.219] W. Erb: Report GSI-P-7-78 (Darmstadt 1978)
[2.220] R. Anholt, W.E. Meyerhof, X. Xu, H. Gould, B. Feinberg, R.J. McDonald, H.E. Wegner, P. Thieberger: Phys. Rev. A **36**, 1586 (1987)
[2.221] R. Anholt, X. Xu, Ch. Stoller, J.D. Molitoris, W.E. Meyerhof, B.S. Rude, R.J. McDonald: Phys. Rev. A **37**, 1105 (1988)
[2.222] Th. Stöhlker, Ch. Kozhuharov, P.H. Mokler, R.E. Olson, Z. Stachura, A. Warczak: J. Phys. B **25**, 4527 (1992)
[2.223] A.S. Schlachter, E.M. Bernstein, M.W. Clark, R.D. DuBois, W.G. Graham, R.M. McFarland, T.J. Morgan, D.W. Mueller, K.R. Stalder, J.W. Stearns, M.P. Stöckli, J.A. Tanis: J. Phys. B **21**, L291 (1988)
[2.224] V.V. Afrosimov, F.F. Barash, A.A. Basalaev, D.B. Korsakov, K.O. Lozhkin, Yu.V. Maidl, V.K. Nikulin, M.N. Panov, L.A. Rassadin, I.Yu. Stepanov: XVIII ICPEAC, Book of Abstracts, p. 600, Canada (1993)
[2.225] L.I. Pivovar, M.T. Novikov, V.M. Tubaev: Sov. Phys. - JETP **15**, 1035 (1962)
[2.226] I.S. Dmitriev, N.F. Vrob'ev, Zh.M. Konovalova: Sov. Phys. - JETP **57**, 1157 (1983)
[2.227] R.D. DuBois: Phys. Rev. A **36**, 2585 (1987)
[2.228] N.V. de Castro-Faria, F.L. Freire, A.G. de Pinho: Phys. Rev. A **37**, 280 (1988)
[2.229] J. Eichler, W.E. Meyerhof: *Relativistic Heavy Ion Collisions* (Academic, New York 1995)
[2.230] R.K. Janev: Nucl. Fusion Suppl. **3**, 71 (1992)

Chapter 3

[3.1] D.F. Dance, M.F.A. Harrison, R.D. Rundel: Proc. Roy. Soc. A **299**, 525 (1967)
[3.2] G.C. Tisone, L.M. Branscomb: Phys. Rev. **170**, 169 (1968)
[3.3] L. Vejby-Christensen, D. Kella, H.B. Petersen, H.T. Schmidt, L.H. Andersen: Phys. Rev. A **53**, 2371 (1996)
[3.4] O. Belly, S.B. Schwartz: J. Phys. B **2**, 159 (1969)
[3.5] B. Peart, D.S. Walton, K.T. Dolder: J. Phys. B **3**, 1346 (1970)
[3.6] L.H. Andersen, D. Mathur, T. Schmidt, L. Vejby-Christensen: Phys. Rev. Lett. **74**, 892 (1995)
[3.7] A.K. Kazansky, K. Taulbjerg: J. Phys. B **29**, 4465 (1996)
[3.8] J.T. Lin, T.F. Jiang, C.D. Lin: J. Phys. B **29**, 6175 (1996)
[3.9] B. Peart, D.S. Walton, K. T. Dolder: J. Phys. B **4**, 88 (1971)
[3.10] P. Defrance, W. Claeys, F. Brouillard: J. Phys. B **15**, 3509 (1982)
[3.11] D.J. Yu, S. Rachafi, J. Jureta, P. Defrance: J. Phys. B **25**, 4593 (1992)
[3.12] C. Bélenger, P. Defrance, E. Salzborn, V.P. Shevelko, H. Tawara, D.B. Uskov: J. Phys. B **30**, 2667 (1997)
[3.13] B. Peart, R. Forrest, K.T. Dolder: J. Phys. B **12**, 847 (1979)
[3.14] B. Peart, R. Forrest, K.T. Dolder: J. Phys. B **12**, L115 (1979)
[3.15] P. Defrance, D.J. Yu, D. Belic: In *XVIII-th Intl Conf. on Physics of Electronic and Atomic Collisions*, ed. by T. Andersen, B. Fastrup, F. Folkmann, H. Knudsen, N. Andersen (AIP, New York 1993) p. 370

[3.16] M. Steidl, D. Hathiramani, G. Hofmann, M. Stenke, R. Völpel, E. Salzborn: In *XIX-th Intl Conf. on Physics of Electronic and Atomic Collisions*, ed. by J.B.A. Mitchell, J.W. McConkey, C.E. Brion (AIP, New York 1995) p.564

[3.17] C. Bélenger, D.J. Yu, P. Defrance: Unpublished (1995)

[3.18] K. Shima, N. Kuno, M. Yamanouchi, H. Tawara: At. Data and Nucl. Data Tables **51**, 173 (1992)

[3.19] H. Tawara, A. Russek: Rev. Mod. Phys. **45**, 178 (1972)

[3.20] H. Tawara: At. Data and Nucl. Data Tables **22**, 491 (1978); IPPJ-AM-1 (Institute of Plasmas Physics, Nagoya University, Nagoya 1977)

[3.21] J.S. Risley: *Electronic and Atomic Collisions* ed. by. N. Oda, K. Takayanagi (North-Holland, Amsterdam 1980) p.619

[3.22] J.S. Risley: Comm. At. Mol. Phys. **12**, 215 (1982)

[3.23] V.I. Radchenko, G.D. Ved'manov: Sov. Phys. - JETP **90**, 1 (1995)

[3.24] I.S. Dmitriev, Ya.A. Teplova, Yu.A. Fainberg: Sov. Phys. - JETP **80**, 28 (1995)

[3.25] T.J. Kvale, J.S. Allen, X.D. Fang, A. Sen, R. Matulioniene: Phys. Rev. A **51**, 1351 (1995)

[3.26] J.S. Allen, X.D. Fang, A. Sen, R. Matulioniene, T.J. Kvale: Phys. Rev. A **52**, 357 (1995)

[3.27] I. Dominiguez, J. de Urquijo, C. Cisneros, H. Martinez, I. Alvarez: Phys. Rev. A **54**, 506 (1996)

[3.28] D.W. Sida: Proc. Phys. Soc. A **68**, 240 (1955)

[3.29] D.R. Bates, J.C. Walker: Proc. Phys. Soc. **90**, 333 (1967)

[3.30] K.L. Bell, A.E. Kingston, P. Madden: J. Phys. B **11**, 3357 (1978)

[3.31] D.P. Dewangan, H.R.J. Walters: J. Phys. B **11**, 3983 (1978)

[3.32] M. Meron, B. Johnson: Phys. Rev. A **41**, 1365 (1990)

[3.33] I.S. Dmitriev, V.S. Nikolaev: Sov. Phys. - JETP **17**, 447 (1963)

[3.34] G.H. Gillespie: Phys. Rev. A **15**, 563 (1977)

[3.35] K. Riesselmann, L.W. Anderson, L. Durand, C.J. Anderson: Phys. Rev. A **43**, 5934 (1991)

[3.36] P. Hvelplund, A. Andersen: Phys. Scr. **26**, 370 (1982)

[3.37] J. Heinemeier, P. Hvelplund, F.R. Simpson: J. Phys. B **9**, 2669 (1976)

[3.38] F. Rahman, I.A. Abbas: Phys. Rev. A **31**, 3974 (1984)

[3.39] B. Hird, I.A. Abbas, M. Bruyere: Phys. Rev. A **33**, 2315 (1986)

[3.40] F. Rahman, B. Hird: At. Data and Nucl. Data Tables **35**, 123 (1986)

[3.41] V. Esaulov: In *Electronic and Atomic Collisions* ed. by. D.C. Lorents, W.E. Weyerhof, J.R. Peterson (North-Holland, Amsterdam 1985) p.175

[3.42] M.M. Duncan, M.G. Menendez: Phys. Rev. A **39**, 1534 (1989); M.M. Duncan, M.G. Menendez, C.B. Mauldin, J.L. Hopkins: Phys. Rev. A **34**, 4657 (1986)

[3.43] F. Penent, J.P. Grouard, J.L. Montmagnon, R.I. Hall: J. Phys. B **25**, 2831 (1992)

[3.44] J. Sørensen, L.H. Andersen, L.B. Nielsen: J. Phys. B **21**, 847 (1988)

[3.45] L. Vikor, L. Sarkadi, F. Penent, A. Bader, J. Palinkas: Phys. Rev. A **54**, 2161 (1996)

[3.46] K.H. Berkner, R.V. Pyle, S.E. Savas, K.R. Stalder: BNL-51304 (Brookhaven National Laboratory, 1980) p. 291

[3.47] F. Melchert, M. Brenner, S. Krüdener, E. Salzborn: Nucl. Instrum. Methods B **99**, 98 (1995)

[3.48] K. Dolder, B. Peart: Rep. Prog. Phys. **48**, 1283 (1985)

[3.49] R.D. Rundel, K.L. Aitken, M.F.A. Harrison: J. Phys. B **2**, 954 (1969)
[3.50] J. Moseley, W. Aberth, J.R. Peterson: Phys. Rev. Lett. **24**, 435 (1970)
[3.51] S. Szücs, M. Karemera, M. Terao, F. Brouillard: J. Phys. B **17**, 1613 (1984)
[3.52] B. Peart, M.A. Bennett, K. Dolder: J. Phys. B **18**, L439 (1985)
[3.53] W. Schön, S. Krüdener, F. Melchert, K. Rinn, M. Wagner, E. Salzborn: J. Phys. B **20**, L759 (1987)
[3.54] D.R. Bates, J.T. Lewis: Proc. Phys. Soc. A **68**, 173 (1955)
[3.55] R. Shingal, B.H. Bransden, D.R. Fowler: J. Phys. B **18**, 2485 (1985)
[3.56] A.M. Ermolaev: J. Phys. B **21**, 81 (1988)
[3.57] V. Sidis, C. Kubach, D. Fussen: Phys. Rev. A **27**, 2431 (1983)
[3.58] F. Borondo, A. Macias, A. Riera: Chem. Phys. Lett. **100**, 63 (1983)
[3.59] L.F. Errea, C. Harel, P. Jimeno, H. Jouin, L. Mendez, A. Riera: Phys. Rev. A **54**, 967 (1996)
[3.60] W. Schön, S Krüdener, F. Melchert, K. Rinn, M. Wagner, E. Salzborn, M. Karemera, S. Szücs, M. Terao, D. Fussen, R.K. Janev, X. Urbain, F. Brouillard: Phys. Rev. Lett. **59**, 1565 (1987)
[3.61] B. Peart, R. Grey, K. Dolder: J. Phys. B **9**, 3047 (1976)
[3.62] F. Brouillard, W. Claeys, G. Poulaert, G. Rahmat, G. van Wassenhove: J. Phys. B **12**, 1253 (1979)
[3.63] B. Peart, R.A. Forrest: J. Phys. B **12**, L23 (1979)
[3.64] G. Poulaert, F. Brouillard, W. Claeys, J.W. McGowan, G. van Wassenhove: J. Phys. B **11**, L671 (1978)
[3.65] T.D. Gaily, M.F.A. Harrison: J. Phys. B **3**, 1098 (1970)
[3.66] B. Peart, R. Grey, K.T. Dolder: J. Phys. B **9**, L373 (1976)
[3.67] B. Peart, D.A. Hayton: J. Phys. B **27**, 2551 (1994)
[3.68] K. Olamba, S. Szücs, J.P. Chenu, N. El Arbi, F. Brouillard: J. Phys. B **29**, 2837 (1996)
[3.69] M. Chibisov, F. Brouillard, J.P. Chenu, M.H. Cherkani, D. Fussen, K. Olamba, S. Szücs: J. Phys. B **30**, 991 (1997)
[3.70] J. Moseley, R.E. Olson, J.R. Peterson: *Case Studies in Atomic Physics* **5**, 1 (North-Holland, Amsterdam 1975)
[3.71] B. Peart, M.A. Bennett: J. Phys. B **19**, L321 (1986)
[3.72] M. Terao, S. Szücs, M. Cherkani, F. Brouillard, R.J. Allen, C. Harel, A. Salin: Europhys. Lett. **1**, 123 (1986)
[3.73] M. Terao, C. Harel, A. Salin, R.J. Allan: Z. Phys. D **7**, 319 (1988)
[3.74] M.H. Cherkani, S. Szücs, H. Hus, M. Terao, F. Brouillard: J. Phys. B **24**, 209 (1991)
[3.75] M.H. Cherkani, S. Szücs, H. Hus, F. Brouillard: J. Phys. B **24**, 2367 (1991)
[3.76] B. Peart, P.M. Wilkins: J. Phys. B **19**, L515 (1986)
[3.77] F. Melchert, W. Debus, M. Liehr, R.E. Olson, E. Salzborn: Europhys. J. **9**, 433 (1989)
[3.78] Y.K. Kim, M. Inokuti: Phys. Rev. A **4**, 665 (1971)
[3.79] F. Melchert, M. Brenner, S. Krüdener, R. Schultze, S. Meuser, K. Huber, E. Salzborn, D.B. Uskov, L.P. Presnyakov: Phys. Rev. Lett. **74**, 888 (1995)
[3.80] L.P. Presnyakov, H. Tawara: Phys. Scr. T **73**, 175 (1997)
[3.81] M.H. Cherkani, D. Fussen, M.I. Chibisov, F. Brouillard: Phys. Rev. A **54**, 1445 (1996)
[3.82] L.P. Presnyakov, H. Tawara, D.B. Uskov: Nucl. Instrum. Methods B **98**, 332 (1995)
[3.83] J.H. McGuire: Adv. At. Mol. Opt. Phys. **29**, 217 (1992)
[3.84] W.H. Aberth, J.R. Peterson: Phys. Rev. A **1**, 158 (1970)

[3.85] J.R. Peterson, W.H. Aberth, J.T. Moseley, J.R. Sheridan: Phys. Rev. A **3**, 1651 (1971)
[3.86] B. Peart, S.J. Foster: J. Phys. B **20**, L691 (1987)
[3.87] B. Peart, S.J. Foster, K. Dolder: J. Phys. B **22**, 1035 (1989)
[3.88] D.A. Hayton, B. Peart: J. Phys. B **28**, L279 (1995)
[3.89] R. Schulze, F. Melchert, M. Hagmann, S. Krüdener, J. Krüger, E. Salzborn, C.O. Reinhold, R.E. Olson: J. Phys. B **24**, L7 (1991)
[3.90] F. Melchert, M. Brenner, S. Krüdener, R. Schultze, S. Meuser, K. Huber, E. Salzborn, D.B. Uskov, L.P. Presnyakov: J. Phys. B **28**, 3299 (1995)
[3.91] D.R. Bates (ed): *Atomic and Molecular Processes* (Academic New York 1962)
[3.92] H. Tawara: Phys. Lett. A **59**, 199 (1976)
[3.93] A.S. Artemov, L.V. Arinin, Yu.K. Baigachev, A.K. Gevorkov: Sov. Phys. - JETP Lett. **53**, 558 (1991)
[3.94] R.F. Holland, D.C. Cobb, W.B. Maier II, W.B. Clodius, P.G. O'Shea, R. Bos, B.C. Frogget: Phys. Rev. A **41**, 2429 (1990)
[3.95] L.H. Andersen, L.B. Nielsen, J. Sørensen: J. Phys. B **21**, 1587 (1988)
[3.96] L.I. Pivovar, Yu.Z. Levchenko: Sov. Phys.-JETP **26**, 27 (1967)
[3.97] D. Rapp, P. Englander-Golden: J. Chem. Phys. **43**, 1464 (1965)
[3.98] H. Tawara, T. Kato: At. Data Nucl. Data Tables **36**, 167 (1987)
[3.99] T.J.M. Zouros: Comm. At. Mol. Opt. Phys. **32**, 291 (1996)
[3.100] E.C. Montenegro, W.E. Meyerhof: Phys. Rev. A **46**, 5506 (1992)
[3.101] A.M. Howard, L.W. Anderson, C.C. Lin: Phys. Rev. Lett. **51**, 2029 (1983)
[3.102] H. Knudsen, J.F. Reading: Phys. Rep. **212**, 107 (1992)
[3.103] L.H. Andersen, P. Hvelplund, H. Knudsen, S.P. Møller, A.H. Sørensen, K. Elsener, K.G. Rensfelt, E. Uggerhøj: Phys. Rev. A **36**, 3612 (1987)
[3.104] L.H. Andersen, P. Hvelplund H. Knudsen, S.P. Møller, J.O.P. Pedersen, S. Tang-Petersen, E. Uggerhøj, K. Elsener, E. Morenzoni: Phys. Rev. A **40**, 7366 (1989)
[3.105] P. Hvelplund, H. Knudsen, U. Mikkelsen, E. Morenzoni, S.P. Møller, E. Uggerhøj, T. Worm: J. Phys. B **27**, 925 (1994)
[3.106] A.L. Ford, J.F. Reading: J. Phys. B **23**, 2567 (1990)
[3.107] A.L. Ford, J.F. Reading: J. Phys. B **27**, 4215 (1994)
[3.108] P.D. Fainstein, V.H. Ponce, R.D. Rivarola: Phys. Rev. A **36**, 3639 (1987)
[3.109] M. Kimura, I. Shimamura, M. Inokuti: Phys. Rev. A **49**, R4281 (1994)
[3.110] R.K. Janev, E.A. Solov'ev, D. Jakimovski: J. Phys. B **28**, L615 (1995)
[3.111] J.M. Hansteen, O.K. Huseby, L. Kocbach: J. Phys. B **26**, L607 (1993)
[3.112] G. Fiorentini, R. Tripoccione: Phys. Rev. A **27**, 737 (1983)
[3.113] S. Cohen, G. Fiorentini: Phys. Rev. A **33**, 1590 (1986)

Chapter 4

[4.1] W. Lotz: J. Opt. Soc. Am. **59**, 915 (1968); *ibid*, **60**, 206 (1970)
[4.2] T.A. Carlson, C.W. Nestor, Jr., N. Wasserman, J.D. McDowell: At. Data **2**, 63 (1970)
[4.3] V. Schmidt: Rep. Progr. Phys. **55**, 1483 (1992)
[4.4] G.N. Marr, J.B. West: At. Data **18**,497 (1976)
[4.5] J.B. West, J. Morton: At. Data **22**, 103 (1978)
[4.6] B.L. Henke, P. Lee, T.J. Tanaka, R.L. Shimabukuro, B.K. Fujikawa: At. Data Nucl. Data Tables **27**, 1 (1982)
[4.7] J.J. Yeh, I. Lindau: At. Data Nucl. Data Tables **32**, 1 (1985)

[4.8] R.E.H. Clark, R.D. Cowan, F.W. Bobrowicz: At. Data Nucl. Data Tables **34**, 415 (1986)
[4.9] E.B. Saloman, J.H. Hubbell, J.H. Scofield: At. Data Tables **38**,1 (1988)
[4.10] G.R. Wight, M.J. Van der Wiel: J. Phys. B **9**, 1319 (1976)
[4.11] D.M.P. Holland, K. Codling, J.B. West, G.V. Marr: J. Phys. B **12**, 2465 (1979)
[4.12] H. Kossmann, V. Schmidt, T. Andersen: Phys. Rev. Lett. **60**, 1266 (1983)
[4.13] N. Saito, I.H. Suzuki: Int. J. Mass. Spectr. and Ion Processes, **115**, 157 (1992)
[4.14] J.M. Bizau, F.J. Wuilleumier: J. Electron. Spectr. Relat. Phenom. **71**, 205 (1995)
[4.15] D.V. Morgan, M. Sagurton, R.J. Bartlett: Phys. Rev. A **55**, 1113 (1997)
[4.16] S.C. Angel, J.A.R. Samson: Phys. Rev. A **38**, 5578 (1988)
[4.17] J.A.R. Samson: Phys. Rev. Lett. **65**, 2861 (1990)
[4.18] S.C. Angel, J.A.R. Samson: Phys. Rev. A **42**, 1307 (1990)
[4.19] B. Rouvellou, L. Journel, J.M. Bizau, D. Cubaynes, F.J. Wuillemier, M. Richter, K.-H. Selbmann, P. Sladeczek, P. Zimmermann: Phys. Rev. A **50**, 4868 (1994)
[4.20] F.J. Wuilleumier, L. Journel, B. Rouvellou, D. Cubaynes, J.-M. Bizau, Z. Liu, J. Liu, M. Richter, P. Sladeczek, K.-H. Selbman, P. Zimmermann: Phys. Rev. Lett. **73**, 3074 (1994)
[4.21] J.B. Donahue, P.A.M. Gram, M.V. Hynes, R.W. Hamm, C.A. Frost, H.C. Bryant, K.B. Butterfield, D.A. Clark, W.W. Smith: Phys. Rev. Lett. **48**, 1538 (1982)
[4.22] Y.K. Bae, J.R. Peterson: Phys. Rev. A **27**, 3265 (1988)
[4.23] T. Koizumi: in *Atomic Physics at High Brilliance Synchrotron Sources.* Proc. Argonne National Lab. Workshop (1994) p. 109 (Report ANL/APS/TM-14, Argonne 1994)
[4.24] B. Sonntag, P. Zimmermann: Rep. Progr. Phys. **55**, 911 (1992)
[4.25] J.H. McGuire: Adv. At. Mol. Opt. Phys. **29**, 217 (1992)
[4.26] J.H. McGuire, N. Berrah, R.J. Bartlett, J.A.R. Samson, J.A. Tanis, C.L. Cocke, A.S. Schlachter: J. Phys. B **28**, 913 (1995)
[4.27] S.L. Carter, H.P. Kelly: Phys. Rev. A **24**, 170 (1981)
[4.28] K. Hino, T. Ishihara, F. Shimizu, N. Toshima, J.H. McGuire: Phys. Rev. A **48**, 1271 (1993)
[4.29] K. Hino, P.M. Bergstrom, Jr., J.H. Macek: Phys. Rev. Lett, **72**, 1620 (1994)
[4.30] A. Dalgarno, H.R. Sadeghpour: Phys. Rev. A **46**, R3591 (1992)
[4.31] J.-Z. Tang, I. Shimamura: Phys. Rev. A **52**, R3413 (1995)
[4.32] A.S. Kheifets, I. Bray: Phys. Rev. A **54**, R995 (1996)
[4.33] K.W. Meyer, C.H. Greene: Phys. Rev. A **50**, R3573 (1994)
[4.34] K.W. Meyer, C.H. Greene, I. Bray: Phys. Rev. A **52**, 1334 (1995)
[4.35] D. Proulx, R. Shakeshaft: Phys. Rev. A **48**, R875 (1993)
[4.36] M. Pont, R. Shakeshaft: Phys. Rev. A **51**, R2676 (1995)
[4.37] L.R. Andersson, J. Burgdörfer: Phys. Rev. Lett. **71**, 50 (1993)
[4.38] M.A. Kornberg, J.E. Miraglia: Phys. Rev. A **48**, 3714 (1993); *ibid* **49**, 5120 (1994)
[4.39] F. Maulbetsch, J.S. Briggs: J. Phys. B **26**, 1679 (1993); *ibid* **26**, L647 (1993); *ibid* **27**, 4095 (1994)
[4.40] C.A. Nicolaides, C. Haritos, Th. Mercouris: Phys. Rev. A **55**, 2830 (1997)
[4.41] A. Dalgarno, J.T. Lewis: Proc. Phys. Soc. (London) A **69**, 285 (1956)

[4.42] A. Huetz, L. Andric, A. Jean, P. Lablanquie, P. Selles, J. Mazeau: In *XIX-th Intl Conf. on Physics of Electronic and Atomic Collisions*, AIP Conf. Proc. **360**, 139-151 (AIP, New York 1995)

[4.43] S.L. Carter, H.P. Kelly: Phys. Rev. A **16**, 1525 (1977)

[4.44] J.D. Jackson: *Classical Electrodynamics*, 2nd ed. (Wiley, New York 1975)

[4.45] W. Heitler: *The Quantum Theory of Radiation*, 3rd edn. (Clarendon, Oxfrod 1954)

[4.46] T. Surić, K. Pisk, R.H. Pratt: in: *X-Ray and Inner-Shell Processes*, 17th Int'l. Conf., Hamburg, Germany (1996), ed. by R.L. Johnson, H. Schmidt-Böcking, B.F. Sonntag, AIP Conf. Proc. **389**, 465 (AIP, New York 1997)

[4.47] O.I. Tolstikhin, D.B. Uskov, V.P. Shevelko: Proc. VIth Int. Conf. Phys. Highly Charged Ions, p. B3 (Kansas State University, Manhattan 1992)

[4.48] R. Moshammer, W. Schmitt, J. Ullrich, H. Kollmus, A. Cassimi, R. Dörner, O. Jagutzki, R. Mann, R.E. Olson, H.T. Prinz, H. Schmidt-Böcking, L. Spielberger: *Ionization of Helium in the Attosecond Equivalent Light Pulse of 1 GeV/u U^{92+} Projectiles*, GSI-Preprint-97-35 (Darmstadt 1997); Phys. Rev. Lett. (1998) (accepted)

[4.49] R.J. Bartlett, P.J. Walsh, Z.X. He, Y. Chung, E.M. Lee, J.A.R. Samson: Phys. Rev. A **46**. 5574 (1992)

[4.50] *Double Photoionization of Helium with Synchrotron X-rays*. Proc. Argonne Physics Devision Workshop (1993) (Report ANL/PHY-94/1, Argonne 1994)

[4.51] *Atomic Physics at High Brilliance Synchrotron Sources*. Proc. Argonne National Lab. Workshop (1994) (Report ANL/APS/TM-14, Argonne 1994)

[4.52] J. Burgdörfer, Y. Qiu, J. Wang, J.H. McGuire: *Double Ionization of Helium by Photons and Charged Particles* in *X-Ray and Inner-Shell Processes*, 17th Int'l. Conf., Hamburg, Germany (1996), ed. by R.L. Johnson, H. Schmidt-Böcking, B.F. Sonntag, AIP Conf. Proc. **389**, 475 (AIP, New York 1997)

[4.53] V. Schmidt, N. Sandner, P.K.H. Dhez, F.J. Wuilleumier, E. Källne: Phys. Rev. A **13**, 1748 (1976)

[4.54] F.J. Wuilleumier: Ann. Phys. (Paris) **4**, 231 (1982)

[4.55] R. Wehlitz, F. Heiser, O. Hemmers, B. Langer, A. Menzel, U. Becker: Phys. Rev. Lett. **67**, 3764 (1991)

[4.56] J.C. Levin, D.W. Lindle, N. Keller, R.D. Miller, Y. Azuma, N. Berrah, H.G. Berry, I.A. Sellin: Phys. Rev. Lett. **67**, 968 (1991)

[4.57] J.C. Levin, I.A. Sellin, B.M. Johnson, D.W. Lindle, R.D. Miller, N. Berrah, Y. Azuma, H.G. Berry, D.-H. Lee: Phys. Rev. A **47**, R16 (1993)

[4.58] N. Berrah, F. Heiser, R. Wehlitz, J. Levin, S.B. Whitfield, J. Viefhaus, I.A. Sellin, U. Becker: Phys. Rev. A **48**, R1733 (1993)

[4.59] F.W. Byron, Jr., C.J. Joachain: Phys. Rev. A **164**, 1 (1967)

[4.60] A. Dalgarno, A.L. Stewart: Proc. Phys. Soc. (London), **76**, 49 (1960)

[4.61] R.C. Forrey, Z.-C. Yan, H.R. Sadeghpour, A. Dalgarno: Phys. Rev. Lett. **78**, 3662 (1997)

[4.62] G.W.F. Drake: In *Long-Range Casimir Forces: Theory and Recent Experiments on Atomic Systems* ed. by F.S. Levin, D.A. Micha (Plenum, New York 1993)

[4.63] Z.-C. Yan, G.W.F. Drake: Chem. Phys. Lett. **259**, 96 (1996)

[4.64] L. Andersson, J. Burgdörfer: Phys. Rev. A **50**, R2810 (1994)

[4.65] H.R. Sadeghpour: Can. J. Phys. **74**, 727 (1996)

[4.66] H. Le Rouzo, C. da Cappollo: Phys. Rev. A **43**, 318 (1991)

[4.67] S. Leonardi, C. Calandra: J. Phys. B **26**, L153 (1993)

[4.68] S.N. Tiwary: J. Phys. B **15**, L323 (1982)
[4.69] M.A. Kornberg, J.E. Miraglia: Phys. Rev. A **48**, 3714 (1993)
[4.70] J.A.R. Samson, Z.X. He. I. Yin, J. Haddad: J. Phys. B **27**, 887 (1994)
[4.71] T.A. Carlson: Phys. Rev. **156**, 142 (1967)
[4.72] J.A.R. Samson, R.J. Bartlett, Z.X. He: Phys. Rev. A **46**, 7277 (1992)
[4.73] T. Kinoshita: Phys. Rev. **105**, 1490 (1957)
[4.74] R.L. Brown: Phys. Rev. A **1**, 586 (1970)
[4.75] T. Åberg: Phys. Rev. A **2**, 1726 (1970)
[4.76] M.Ya. Amusia, E.G. Drukarev, V.G. Gorshkov, M.P. Kazachkov: J. Phys. B **8**, 1248 (1975)
[4.77] T. Ishihara, K. Hino, J.H. McGuire: Phys. Rev. A **44**, R6980 (1991)
[4.78] R.C. Forrey, H.R. Sadeghpour, J.D. Baker, J.D. Morgan III, A. Dalgarno: Phys. Rev. A **51**, 2112 (1995)
[4.79] J.C. Levin, G.B. Armen, I.A. Sellin: Phys. Rev. Lett. **76**, 1220 (1996)
[4.80] R. Dörner, T. Vogt, V. Mergel, H. Khemliche, S. Kravis, C.L. Cocke, J. Ullrich, M. Unverzagt, L. Spielberger, M. Damrau, O. Jakutzki, I. Ali, B. Waever, K. Ullmann, C.C. Hsu, M. Jung, E.P. Kanter, B. Sonntag, M.H. Prior, E. Rotenberg, J. Denlinger, T. Warwick, S.T. Manson, H. Schmidt-Böcking: Phys. Rev. Lett. **76**, 2654 (1996)
[4.81] G.H. Wannier: Phys. Rev. **90**, 817 (1953)
[4.82] U. Fano: Rep. Progr. Phys. **46**, 97 (1983)
[4.83] M.S. Lubell: Z. Phys. D **30**, 79 (1994)
[4.84] J.F. McCann, D.S.F. Crothers: J. Phys. B **19**, L399 (1986)
[4.85] P.K. Kabir, E.E. Salpeter: Phys. Rev. **108**, 1256 (1957)
[4.86] J.A.R. Samson, C.H. Greene, R.J. Bartlett: Phys. Rev. Lett. **71**, 201 (1993)
[4.87] H.R. Sadeghpour, A. Dalgarno: Phys. Rev. A **47**, R2458 (1993)
[4.88] C. Bélenger, P. Defrance, E. Salzborn, V.P. Shevelko, H. Tawara, D.B. Uskov: J. Phys. B **30**, 2667 (1997)
[4.89] L. Andersson, J. Burgdörfer: in *Double Photoionization of Helium with Synchrotron X-rays.* Proc. Argonne Physics Division Workshop (1993) (Report ANL/PHY-94/1, Argonne 1994) p. 91
[4.90] L. Spielberger, O. Jagutzki, R. Dörner, J. Ullrich, U. Meyer, V. Mergel, M. Unverzagt, M. Damrau, T. Vogt, I. Ali, Kh. Khayyat, D. Bahr, H.G. Schmidt, R. Fraham, H. Scmidt-Böcking: Phys. Rev. Lett. **74**, 4615 (1995)
[4.91] D. Morgan, M. Sagurton, R.J. Bartlett: Phys. Rev. A **55**, 1113 (1997)
[4.92] T. Surić, K. Pisk, B.A. Logan, P.H. Pratt: Phys. Rev. Lett. **73**, 790 (1994)
[4.93] M. Sagurton, R.J. Bartlett, J.A.R. Samson, Z.X. He, D. Morgan: Phys. Rev. A **52**, 2829 (1995)
[4.94] M. Ya. Amusia: In *Double Photoionization of Helium with Synchrotron X-rays.* Proc. Argonne Physics Division Workshop (1993) (Report ANL/PHY-94/1, Argonne 1994) p. 79
[4.95] M. Ya. Amusia, A.I.Mikhailov: Phys. Lett. A **199**, 209 (1995)
[4.96] M. Ya. Amusia, A.I.Mikhailov: J. Phys. B **28**, 1723 (1995)
[4.97] P.M. Bergstrom Jr., K. Hino, J.H. Macek: Phys. Rev. A **51**, 3044 (1995)
[4.98] T. Surić, K. Pisk, P.H. Pratt: Phys. Lett. A **211**, 289 (1995)
[4.99] V.E. Brijunas, A.V. Kupliauskienė, Z.J. Kupliauskis: Lietuvos fizikos rinkinys **24**, 92 (1984)
[4.100] O.I. Zatsarinny, L.A. Bandurina: J. Phys. B **26**, 3765 (1993)
[4.101] T. Brage, Ch. Froese-Fischer, N. Vaeck: J. Phys. B **26**, 621 (1993)
[4.102] C. Nicolaides, D. Beck: Can. J. Phys. **53**, 1224 (1975)

[4.103] Z.J. Kupliauskis: Izvestiya Akademii Nauk SSSR (ser. Fizicheskaya) **41**, 2626 (1977)
[4.104] A.S. Kheifets: J. Phys. B **27**, L463 (1994)
[4.105] H.W. Wolf, K. Radler, B. Sonntag, R. Haensel: Z. Phys. **257**, 353 (1972)
[4.106] R.D. Drives: J. Phys. B **9**, 817 (1976)
[4.107] T.A. Ferett, D.W. Lindle, P.R. Heimann, W.D. Brewer, U. Becker, H.G. Kerkhoff, D.A. Shirley: Phys. Rev A **36**, 3172 (1987)
[4.108] D. Cubaynes, J.M. Bizau, F.J. Wuilleumier, B. Carré, F. Gounand: Phys. Rev. Lett. **63**, 2460 (1989)
[4.109] M. Richter, J.M. Bizau, D. Cubaynes, T. Mengel, F.I. Wuilleumier, B. Carre: Europhys. Lett. **12**, 35 (1990)
[4.110] D. Cubaynes, J.M. Bizau, M. Richter, F.J. Wuilleumier: Europhys. Lett. **14**, 747 (1991)
[4.111] D. Cubaynes, J.M. Bizau, T.J. Morgan, B. Carré, F.J. Wuilleumier: Presented at 10th Intl Conf. on VUV Radiation Physics, ed. by Y. Petrov (Paris 1992) p. Tu79
[4.112] D. Cubaynes: Unpublished (1993)
[4.113] F. Combet Farnoux, M. Lamoureux: J. Phys. B **9**, 897 (1976)
[4.114] C.E. Theodosiou, W. Fielder, Jr.: J. Phys. B **15**, 4113 (1982)
[4.115] A. Lisini, P.G. Burke, A. Hibbert: J. Phys. B **23**, 3767 (1990)
[4.116] G.B. Armen, F.P. Larkins: J. Phys. B **24**, 2675 (1991)
[4.117] Z. Felfli, S.T. Manson: Phys. Rev. Lett. **68**, 1687 (1992)
[4.118] A. Kupliauskienė: Phys. Scr. **53**, 149 (1996)
[4.119] A. Kupliauskienė: Phys. Scr. **55**, 445 (1997); J. Phys. B **30**, 1865 (1997); J. Phys. B **31**, 2885 (1998)
[4.120] V.L. Jacobs, P.G. Burke: J. Phys. B **5**, L67 (1972)
[4.121] J.A. Richards, F.P. Larkins: J. Electron Spectrosc. Relat. Phenomen. **32**, 193 (1983)
[4.122] S. Salomonson, C.L. Carter, H.P. Kelly: Phys. Rev. A **39**, 5111 (1989)
[4.123] K.A. Berrington, P.G. Burke, W.C. Fon, K.T. Taylor: J. Phys. B **15**, L603 (1980)
[4.124] T.N. Chang: J. Phys. B **13**, L551 (1980)

Index

Printing: Mercedesdruck, Berlin
Binding: Buchbinderei Lüderitz & Bauer, Berlin